Early Power and Transport

Young Engineer's Guide to Various and Ingenious Machines

Bryan Lawton

Portions Reprinted from *Various and Ingenious Machines*, published by Brill, Copyright 2004 (with permission). This edition published in 2017 by ASME Press (The American Society of Mechanical Engineers, Two Park Avenue, New York, NY 10016, USA, www.asme.org).

All rights reserved. Printed in the United States of America. Except as permitted under the United States Copyright Act of 1976, no part of this publication may be reproduced or distributed in any form or by any means, or stored in a database or retrieval system, without the prior written permission of the publisher.

INFORMATION CONTAINED IN THIS WORK HAS BEEN OBTAINED BY THE AMERICAN SOCIETY OF MECHANICAL ENGINEERS FROM SOURCES BELIEVED TO BE RELIABLE. HOWEVER, NEITHER ASME NOR ITS AUTHORS OR EDITORS GUARANTEE THE ACCURACY OR COMPLETENESS OF ANY INFORMATION PUBLISHED IN THIS WORK. NEITHER ASME NOR ITS AUTHORS AND EDITORS SHALL BE RESPONSIBLE FOR ANY ERRORS, OMISSIONS, OR DAMAGES ARISING OUT OF THE USE OF THIS INFORMATION. THE WORK IS PUBLISHED WITH THE UNDERSTANDING THAT ASME AND ITS AUTHORS AND EDITORS ARE SUPPLYING INFORMATION BUT ARE NOT ATTEMPTING TO RENDER ENGINEERING OR OTHER PROFESSIONAL SERVICES. IF SUCH ENGINEERING OR PROFESSIONAL SERVICES ARE REQUIRED, THE ASSISTANCE OF AN APPROPRIATE PROFESSIONAL SHOULD BE SOUGHT.

ASME shall not be responsible for statements or opinions advanced in papers or ... printed in its publications (B7.1.3). Statement from the Bylaws.

For authorization to photocopy material for internal or personal use under those circumstances not falling within the fair use provisions of the Copyright Act, contact the Copyright Clearance Center (CCC), 222 Rosewood Drive, Danvers, MA 01923, Tel: 978-750-8400, www.copyright.com.

Requests for special permission or bulk reproduction should be addressed to the ASME Publishing Department, or submitted online at: https://www.asme.org/shop/books/book-proposals/permissions

ASME Press books are available at special quantity discounts to use as premiums or for use in corporate training programs. For more information, contact Special Sales at CustomerCare@asme.org

Library of Congress Cataloging-in-Publication Data

Names: Lawton, B. (Bryan) author.
Title: Early power and transport : young engineer's guide to various and
  ingenious machines / Bryan Lawton.
Description: New York, NY, USA : ASME Press, 2017. | Includes bibliographical
  references and index.
Identifiers: LCCN 2017042509 | ISBN 9780791861417
Subjects: LCSH: Power resources--History. | Transportation--History. |
  Conveying machinery--History.
Classification: LCC TJ163.2 .L374 2017 | DDC 621.8/12--dc23
LC record available at https://lccn.loc.gov/2017042509

For my grandsons, Tom and Kit.

Table of Contents

Foreword	vii
Preface	ix

1. Animal Power — 1
   Introduction — 1
   Man Power — 3
   Slavery — 7
   Animal Power — 9
   Appendices — 15
      Treadmills — 16
      Muscular Activity in Animals — 19

2. Water Power — 23
   Introduction — 23
   Vertical-Axis (Norse or Greek) Mills — 23
   Horizontal-Axis (Roman) Mills — 29
   Appendices — 41
      Power Output of Water Wheels — 41
      Horizontal-Axis Undershot Wheels (Roman Mills) — 42
      Vertical-Axis Wheels (Greek or Norse Mills) and Pelton's Wheel. — 46

3. Wind Power — 51
   Introduction — 51
   Vertical-axis Windmills — 51
   Horizontal-axis Windmills — 53
   Appendices — 67
      Theoretical Power of Windmills — 67
      Wind Distribution — 70

4. Land Transport — 75
   Introduction — 75
   Prehistory and the Ancient Civilisations — 75
   Riding — 80
   Greece and Rome — 81
   Medieval, Renaissance, and Early Modern — 85
   Early Modern Period — 87
   Appendices — 95
      Two-wheeled Vehicles — 95
      Forces on Draught Beasts: Limits of Traction — 98
      Wheels on Soft Surfaces — 101
      Wheels on Hard Surfaces — 106

		Power and Maximum Velocity	109
		Steering	110
5.	**Water Transport**		115
	Introduction		115
	Ancient Civilisations		115
	Greece and Rome		121
	Medieval and Renaissance		125
	Early Modern		132
	Appendices		138
		Force and Power Required by Ships	138
		Strength of Hulls	143
		Rowing	146
		Sailing	150
6.	**Undersea and Aerial Transport**		153
	Introduction		153
	Undersea Craft		153
	Air Craft		159

BIBLIOGRAPHY 165

AUTHOR INDEX 171

SUBJECT INDEX 175

Foreword

Bryan Lawton is the author of the acclaimed *Various and Ingenious Machines*. VIM is a 1300 page, two-volume work, which provides a well-researched history of mechanical engineering worldwide, from prehistory to roughly the beginnings of industrialization. The author received the American Society of Mechanical Engineering's Engineering-Historian Award for VIM, in 2016.

Dr. Lawton is a retired Reader in Thermal Power at Cranfield University in Shrivenham, UK. One of the main purposes of his two-volume VIM is to not only deliver an outline of mechanical engineering history, but also to provide an adequate mathematical treatment for each topic, without straying beyond an engineering first degree level. Thus the reader can enjoy a well-written history and can supplement it with a concise mathematical explanation in the appendices at the end of most of the chapters.

The author has told me that VIM represents about ten years of research and three years of solid writing. The result is a seminal work on the historical mechanical engineering theme, "If it moves, it's mechanical."

It is here, we are now treated by Dr. Lawton to a *vade mecum* of VIM, the *Young Engineer's Guide to Various and Ingenious Machines*. What follows this Foreword is a guidebook to the magisterial 1300 page VIM, with shortened chapters but still containing elements of the mathematical appendices.

Thus we have VIM in a shortened, more affordable and accessible form, to serve as a guide for possibly probing deeper into the parent work. In this Volume 1, we have six chapters on Power and Transportation which provide a synopsis of the thirteen chapters in VIM's Volume 1. Briefly they are:

- Chapter 1 deals with muscular work and power, from both humans and animals. Even as late as 1850, 94% of America's industrial power was supplied by humans and animal muscle, with machines supplying the remaining 6%.
- Waterpower, the subject of Chapter 2, is shown to have origins going back to early Greek times, even chronicled in poetry.
- Chapter 3 deals with wind power, tracing the development of windmills which appear some 700 years after the first use of water power.
- Land transport is dealt with in Chapter 4, covering the development of wheeled vehicles.
- Chapter 5 covers the history of transport by water, with a history of sailing ships.
- Balloons and submarines are subjects in Chapter 6 on undersea and aerial transport.

In these chapters, the reader will be rewarded with appendices giving mathematical explanations to supplement the written material.

Lee S. Langston
ASME Life Fellow
Professor Emeritus
Mechanical Engineering
University of Connecticut
Storrs, CT 06269

Preface

This volume grew out of "Various and Ingenious Machines: The Early History of Mechanical Engineering", which was first published in 2004 in two volumes. It was intended to cover the whole of early mechanical engineering history from Palaeolithic flint mining through to the start of the industrial revolution. It was quite extensive, but in 2016, Lee S. Langston, Professor Emeritus, University of Connecticut, suggested that some key chapters from the original might be condensed to produce something shorter and less expensive. He thought that engineering students, many of whom are very interested in the history of mechanical devices and how they work, might benefit.

Thus, the task was to reduce the 1274 pages of the original to two volumes, each of about 150 pages. This could be achieved only by selecting key sections of the original and condensing them without losing too much of their impact. Consequently, this volume contains six chapters on power generation and transport during the ancient, classical, medieval, and early modern periods. A sister volume covers mining and metalworking.

Each chapter closes with a short appendix covering the elementary mathematical theory, up to first degree level. Such analysis should be of interest to engineering students and may also be useful to experimental archaeologists and others who, not being content simply to read about the alleged performance of historical machines, re-create and test them. And if the reader prefers, he can simply omit the appendices without losing the historical development of the subject.

Thanks are due to Professor Lee S. Langston, for his suggestion, mentioned above, and for generously agreeing to write a Foreword. I am also grateful to the original publishers, E J Brill, Leiden, for permission to make this selection. Finally, I must thank my wife, Barbara, for her careful checking of the manuscript, for many useful suggestions, and for her great forbearance.

Bryan Lawton.

Chapter 1 Animal Power

Introduction

Throughout the pre-industrial period, the energy supplied to mechanical devices was obtained mostly from muscle, and a short summary of the power and force available through human and animal power is shown in Table 1.1. For short periods of time these values may be increased by perhaps four times. Apart from the water mill and windmill, animal or human labour invariably drove machinery. Even weight driven machinery, such as clocks and ballistae, required men or animals to raise their weights.

Table 1.1 Power output of man and animals.
After Rankine(1889, 250-55)

	Force (N)	Velocity (m/s)	Power (W)
Draft Horse	540	1.2	650
Ox	540	0.8	430
Mule	270	1.2	325
Donkey	130	1.2	155
Man pumping	60	0.8	48
Man turning winch	80	0.8	64

It was not until comparatively late that the thoughts of educated men turned to this difficulty. The Minutes of the Royal Society in 1670, for example, record that the most famous mechanic of the age, namely their Secretary, Robert Hooke, examined the problem, see Jennings (1995,15). "[He] produced a contrivance of his to try, whether a mechanical muscle could be made by art, performing without labour the same office which a natural muscle doth in animals". However, Hooke's normally fertile and inventive brain let him down on this occasion and he "was ordered to consider more fully of it". It was not until the 20th century that significant progress was made, and described by Hill (1935, 353-81), one of Hooke's successors at the Royal Society, who gave the Thomas Hawksley Lecture to the Institution of Mechanical Engineers in 1935. The performance of muscles has remained a pertinent consideration and one of great interest to engineers. In the field of bioengineering, for example, the engineer seeks to control and enhance muscular activity in injured or diseased patients, and in modern robotics, the computer-controlled hydraulic ram tirelessly fulfils the functions of the muscle, succeeding where Robert Hooke failed.

Atkinson (1960, 51) has summarised the power sources that were available during the immediate pre-industrial period. He took for his data the illustrations used and discussed by five authors during the period 1556 to 1772, namely: Agricola (1556), Ramelli (1558), Zonca (1607), Leupold (1735) and Diderot (1772). Between them, they described and illustrated 354 power driven machines, but even so, it is questionable if these machines can be taken as a fair sample of the power sources then in use. Their data, however, seem reasonable and consistent and may be accepted in the absence of anything better, see Table 1.2.

Table 1.2 Analysis of Pre-Industrial Power Sources

Source of Power	Agricola 1556	Ramelli 1588	Zonca 1607	Leupold 1735	Diderot 1772	Total
Man	24 52%	89 55%	6 24%	4 16%	35 37%	158 45%
Water	17 37%	65 40%	10 40%	13 52%	33 34%	138 39%
Animal	4 9%	5 3%	7 28%	6 24%	22 23%	44 12%
Other	1 2%	3 2%	2 8%	2 8%	6 6%	14 4%
Total	46	162	25	25	96	354

Some 45% of these machines were driven by manpower, usually through a crank or a treadmill. The use of manpower declined during this period from about 55% of applications in the 16th century to around 30% in the 18th century. Water was the next most common source of power and was used in about 40% of applications and remained relatively constant throughout the period. Waterwheels were usually horizontal-axis machines that were undershot, overshot, or of breast wheel configuration, although some vertical-axis machines were also illustrated. Animal power, usually derived from horses or oxen, was illustrated in 12 % of applications but this increased from about 6% in the 16th century to about 24% in the 18th century. Other sources of power for these machines, mainly wind-power or gravity, amounted only to about 4%, and increased from 2% in the 16th century to about 7% in the 18th century. The proportion for wind-power seems rather low but probably reflects the unreliable nature of wind and the restrictions it placed on machine location.

It appears, therefore, that the use of water mills remained steady and that the use of manpower slowly declined and was replaced by an increasing appliance of animal power and, to a lesser extent, of wind power. These data refer to power sources for machines such as pumps, oil mills, fulling mills, and corn mills; they do not include: the near total reliance of land transport on horses, oxen, and walking; of water transport on man (rowing) and wind power (sails); or the military use of gunpowder. The muscles of men and animals, as Table 1.2 suggests, remained an important source of power well into the industrial period and placed severe restrictions on the wealth that could be generated. Even as late as 1850 machines supplied only 6% of America's industrial power, animals supplied 79%, and human muscle supplied 15% (Scientific American, September 2001, p11).

Man Power

It need hardly be said that the power of his own muscle was the first source of energy available to man and throughout history illustrations showing men exerting themselves are not too difficult to find. Pushing, pulling, striking, rotating, lifting, or climbing are commonly used and may be undertaken either stationary or when moving relative to the earth.

Figure 1.1 portrays the massed use of manpower in Assyria to move a bull-headed statue for the palace of Sennacherib (c704-681 BC). The construction of such colossi seems to have had a great appeal to ancient civilisations. We think of the great stone circles of Stonehenge, Avebury and Brittany, the Colossi of Memnon, the Pyramids, and the other Seven Wonders of the Ancient World. A man can lift more than his own body-weight, about 700 N, but only for a few seconds. If he has a firm foothold and pulls a rope whilst leaning backwards, he can exert a substantial proportion of his body weight, typically about 450 N, for a comparatively short time. For long term exertions, Rankine (1889, 250-55) recommends that a pulling force of 120 N (27 lb) for periods up to eight hours may be acceptable.

Lubricated hardwood, sliding across the grain of the same hardwood, has a coefficient of friction of only 0.072 so a 1 tonne load

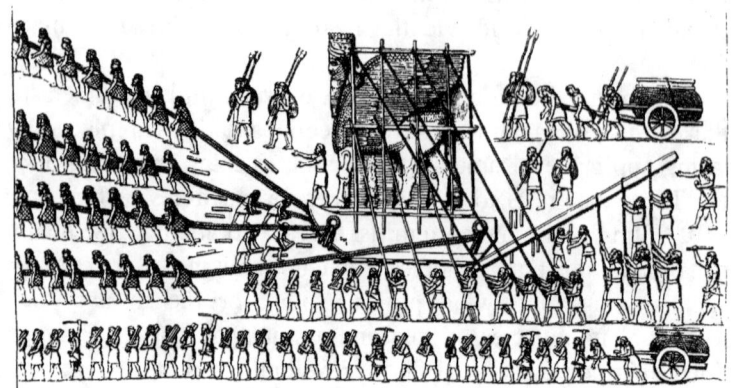

Figure 1.1 Manpower used to lever and pull a large bull-headed statue for the palace of Sennacherib (c704-681 BC).

requires a horizontal pull of 706 N. Assuming a man pulls 120 N then the number of men required per tonne is six, over horizontal ground, and even more to ascend a slope.

Another example of muscular lifting or pushing is in the operation of bellows. These are used to increase the air supply rate to a fire and thus increase its temperature. They were commonly used in smelting and in the blacksmith's forge to heat metal to the necessary elevated temperatures. Without them many metalworking processes would have been very difficult or even impossible. The Catalan furnace, used to smelt iron during the Roman period, required the use of two bellows blowing alternately to give a continuous blast. Many devices were used to lift bellows; Biringuccio (1540, 304) described that shown in Figure 1.2. "It is also customary," he says "especially for master founders, to make the bellows move by tying a piece of rope to a scaffolding or to something else that is above the bellows so that it comes between them. To this is tied a transverse piece of wood which such masters usually call a yoke. Standing alternately first on one and then on the others bellows, a man causes it to make a blast by bearing down on it with his weight." The yoke is fixed in the centre so as one bellows falls the other must rise. The man lifts his own weight from the lower bellows to the higher bellows to make them work and is more akin to climbing, perhaps, than to lifting.

Figure 1.3 (top) shows a Roman treadmill operated by slaves and

Figure 1.2 Bellows operated by a man's weight. Biringuccio (1540, 304).

used to lift heavy weights, and the bottom image illustrates one operated by prisoners, holding onto a stationary bar as they climb up a slowly rotating stepped wheel. Men on the inside are in stable equilibrium, but those on the outside are unstable and need to hold onto a bar. It is demonstrated in the Appendix that treadmills have the peculiar characteristic that an increase in the number people on the wheel does not reduce the work, by sharing it amongst a larger number, but increases the work demanded of everyone. This is because an increase in the weight on the treadmill increases the wheel's speed and hence increases the climbing rate and the work output of everyone on the wheel. One could be cruel whilst appearing to be kind. In the prisons, treadmills often did no useful work but simply drove a brake to provide a punishment. The speed of the mill could be adjusted by a screw mechanism on the brake and for this reason the prison warder was often referred to, then as now, as the "screw." Tightening the screw slowed the treadmill and hence made it easier for the prisoners.

Figure 1.3 Top: slaves operating a large treadmill crane. Tomb of the Haterii. c100 AD. Museo Lateranese, Rome. Bottom: Horizontal axis treadmill operated by ten men at Brixton Prison, London, about 1800.

Such prison treadmills continued into the early 20th century. Haden (1928) in a letter to the Board of Education reports that his company had supplied prisons in Salford, Hong Kong, St Vincent, Natal, Antigua, Sierra Leone etc. The largest wheels were for Salford Gaol where 80 men operated five treadmills on two floors. The power was used in pumping water from wells for prison use. At Devizes Gaol, it was used for grinding corn, at York for polishing marble, and several colonial wheels were used to power stone crushers. Haden says the wheels were 1.73 m diameter and when stepping at 28 treads per minute the wheel made 1.8 revolutions per minute. This implies a climbing rate of 587 m/hour. This was adequate for most purposes. The men worked the wheel for 20 minutes at a time and a brake stopped rotation during changeover periods. Haden thought it essential that the speed of the wheel should be kept practically constant, to prevent overworking the men, and consequently all wheels were fitted with a governor to regulate the speed.

Slavery

The practice of slavery probably dates to prehistoric times, but it is unlikely to have become customary (institutionalised) until the Neolithic Period when improved agriculture enabled more complex societies to evolve and we have definite evidence for warfare. Agricultural surplus allowed specialist occupations to develop, including, of course, warriors, who defended agrarian community against the depredations of nomadic hunter-gatherers. The latter, no doubt, were unaware that they were not supposed to gather what others had planted. Certainly, Neolithic cave paintings frequently illustrate warfare (archers in combat) and even execution scenes, see Kühn (1952, plate XXVII, XXXII, XXXIV). Before industrialisation, a large part of the drudgery of manual work was undertaken by slaves, and in particular, by women. Most commonly such slaves were obtained either through raiding, or conquests of neighbouring peoples, or later, from within the society itself when some sold themselves or their family to pay debts, or were enslaved for crimes. Presumably, slavery became possible when the value of a captive's labour was worth more than his keep, that is, when a society had advanced to the stage of producing an agricultural surplus.

Slaves and free men worked side by side but the number of slaves in any society was usually, but not invariably, less than the number of free workers. They were a possible source of revolt and their existence perhaps reduced the price of free labour. More importantly, the slaves lowered the social standing of their free colleagues. The existence of slavery should, perhaps, have acted as an incentive to develop labour saving machines and to improve ancient engineering, but it did not. If slavery inhibited technological progress, it was because intelligent and able men were not attracted to the improvement of practical work having a low social position.

The sale of a slave is illustrated in Figure 1.4, which is taken from a Capuan funeral stele, of the 1st century AD. A nude man, the slave, stands on a stone. On the right, a man in a Roman toga extends his right arm towards the slave, evidently asking the price. On the left a man wearing a Greek chiton, the seller, points out the slave's finer features. Manifestly, the deceased was a great man in Capua who nevertheless wished to record his humble beginnings, as a slave, on his tombstone.

Figure 1.4 Sale of a slave in Capua, 1st C AD.

Slavery was an accepted feature, and was an essential economic attribute, of all pre-industrial civilisations. Without slaves to undertake the more mundane and burdensome tasks, an upper class of citizens, it was argued, would not be free to attend to the business of the state. The universal rules of slavery, that helped to make it acceptable, were that slaves should not be treated cruelly and should not be of one's own race, society or religion unless they were criminals or debtors. It was so tenacious that religious convictions did little to reduce the problem, notwithstanding Christianity's early popularity with slaves. It was not banned until the nineteenth century, by which time industrialisation and the development of mechanical machines had greatly increased the productivity of labour to the point where such a ban was economically feasible.

Animal Power

The modes of muscular work that were applied to manual activity may also be applied to animals, that is, muscular work may be due to pulling, pushing, striking, rotating, lifting, or climbing and be undertaken either stationary or moving. Applied to animals it is immediately apparent that although stronger or faster than humans, they are much less adaptable. They are not used in pushing, except very occasionally as in a reaper used by the Romans, and they are not used in striking, or in rotary motions having a small radius. In lifting, they are used as pack animals. The main application of animal power is, in fact, to moving and pulling, that is, to the transport of heavy loads or to drive a capstan (whim or gin). However, in these applications they have been invaluable.

Horses and asses were used increasingly during second millennium BC and the first and second millennia BC was the great era of the chariot. The slow, heavy chariots depicted in Figures 4.2 and 4.4 (see chapter 4) were replaced by lighter and faster designs, Figure 4.3. In the 5th century BC rotary mills driven by donkeys first made their appearance and about 300 BC they were used in Athens for grinding corn, Humphrey et al (1999, 35). Horses were commonly used in warfare, transport, and agriculture but, unlike oxen, their use was limited by the design of their harness and they were unable to exert their full strength. Gimpell (1993, 32) points out that the Theodosian Code of 438 AD decreed that the maximum load of carts etc was limited to 500 kg. Probably this was to prevent damage to the road surface, but Lefebvre des Noëttes (1931), thought that the ox-harness, which had been in use since prehistoric times, was applied without modification, to horses, donkeys, and mules. The result was that they could not pull heavy loads without choking. Figure 1.5 (top) shows an ancient harness with neck and body girths and the neck girth presses on the horse's windpipe when it attempts to pull a heavy load. Lefebvre des Noëttes

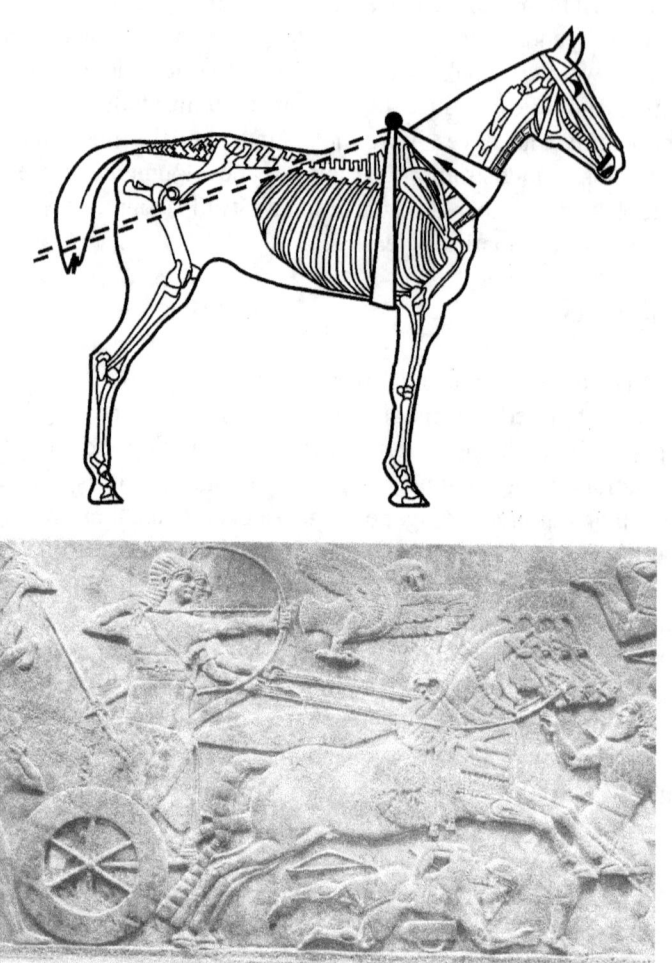

Figure 1.5 Top: ancient horse harness with neck and body girths. When the horse pulls, it constricts its windpipe. After Klemm (1959, 82).
Bottom: 9th century BC chariot of Ashurnasirpal trampling his enemies. The prancing pose is dramatic but may be due to constriction of the windpipe.

carried out experiments using modern copies of ancient harness and found that horses had difficulty in pulling loads more than about 500 kg. Forbes (1955, II, 82) estimates that, instead of pulling fifteen times that of a man, the horse in antiquity barely pulled four times this amount.

The ancient harness was inefficient because the point of attachment was at the back of the horse's neck and the flexible collar or band was set high on the throat instead of a rigid collar resting on the shoulders as in the modern harness. This design appears on numerous ancient wall paintings and bas-reliefs, Figure 1.5 (bottom), and on many Greek vases depicting ancient myths. The horse, when galloping, is usually shown prancing and with its head thrown back. Although the pose is dramatic, the horse may be reacting against the pressure on its windpipe. However, it seems inconceivable that, for hundreds of years, no one noticed that the design of harness choked the horse when pulling heavy loads. Recent opinion (Adams, 2012, 218-240) considers Lefebrvre's work to be largely discredited, despite much distinguished support.

Another reason for the comparatively late adoption of the horse in heavy pulling is that it was more expensive than the ox, and this, perhaps, was the reason no one bothered to design a better harness until about 1000 AD. Hassall (1973, 128) quotes a section from the thirteenth century English "Book of Husbandry" that discusses the relative merits of horse plough and ox plough and comes down firmly in favour of the ox plough, even though the horse harness then in use was satisfactory. "The plough of oxen is better than the plough of horse but if it be upon stony ground it grieveth sore the oxen in their feet. And the plough of horse is more costly than the plough of oxen and yet shall your plough of oxen do as much work as your plough of horse though you drive your horse faster than you do your plough of oxen yet in what ground so ever it be your plough of oxen if you till your land well and evenly they shall do as much work one day as with another as your plough of horse and if the ground be tough your oxen shall work where your horse shall stand still." This last remark suggests that even in the 13th century the horse remained restricted by its harness although better horse collars had been devised.

The estimated cost of a horse with feed, shoeing, straw, hay etc was 9s 6½ d (48p) per year, whereas an ox costs cost a third this, 3s 1d (15.5p) per year. "And if a horse be overset and brought down with labour it is adventure and ever he recover it. And if your ox be overset and brought down with labour you shall for 12d [5p] in summer season have him so pastured that he will be strong enough to do your work or else shall be so fat that you may sell him for as much money as he cost you."

The proper use of the horse was one of the great technical achievements of the late Middle Ages. This was made possible by the invention of an improved harness, having a lower attachment point. The earliest depiction of a horse using the improved harness is in the Bayeux Tapestry of the 11th century. This shows a horse pulling a harrow. A rigid collar resting on the horse's shoulders enabled it to exert maximum effort without danger. A further invention, or re-invention, of the period was the iron horseshoe. The horse's natural homeland is the grasslands of the steppes and on rocky terrain, or metalled roads, their hooves were subjected to wear and damage. Nailed onto their hooves, the horseshoe gave the necessary protection.

The use of oxen continued, in agriculture and in heavy transport, until the early part of the 20th century. Figure 1.6, for example, shows 86 oxen towing a post mill on 28th March 1797 when it was necessary to

Figure 1.6 Oxen towing a windmill in 1797. Preston Manor, Brighton and Hove.

move a mill from Regency Square to the Dyke Road, Brighton; about two miles. Many of the local farmers contributed one of more oxen and a substantial proportion of the local populace turned out to watch the event. Windmills were often taken down and re-erected on a new site, although not, perhaps, in such a grand manner. Rankine (1889, 251) says that the force exerted by an ox is 120 lb (535 N) consequently the 86 oxen could exert a pull of 10,320 lb (4.68 tonnes) if all pulled at the same time. The coefficient of friction of wood sliding over grass is about 0.75

consequently the greatest load that could be pulled is 6.24 tonnes. Of course, had the windmill been mounted on wheels the coefficient of rolling friction would be quite low and fewer beasts would be required. It was easier, apparently, to yoke 86 oxen than to fit wheels.

Atkinson (1960, 31-55) thought that the earliest animal-powered rotary machine was probably the Roman corn mill of about 200 BC, but it was not until the 11th or 12th century that the adoption of an improved horse collar permitted the horse to be an effective source of power. Early experiments with horse driven mills, however, were not always successful. Bennett and Elton (1898, 195) record the problems experienced at Dunstable Abbey in 1295. "This year brother John, the carpenter, made a new mill constructed upon principles hitherto unknown, promising that one horse should be able to turn it; but when it was made and should have ground, four strong horses could scarcely move it; and so it was removed and the old horse mill resumed." Here the fault seems to lie with the design of the mill rather than the horse harness. However, by the 16th century some very large horse-driven machines were in use and were described by Agricola (1556,167), Ramelli (1588, 311) and Zonca (1607). Figure 1.7 shows a heavy horse in a well-designed harness driving a corn mill through various pairs of gears designed to increase the rotational speed of the stones. The horse is tethered at the collar to ensure it walks in a circle. In other designs, Ramelli has the horse walk inside a circular wall or fence.

The arrival of the railways and steam locomotives prompted the invention of a locomotive powered by four horses walking on a treadmill. The *Impulsoria*, a drawing of which appeared in the *Illustrated London News* in 1850, was an Italian invention and was promoted mainly for its economy, but it failed for it was never a serious competitor of the steam locomotive. Indeed, at the famous Rainhill trials, in 1829, the steam locomotive completely outclassed the horse-driven *Cyclopede* that was exhibited there.

It was not only the ox and the horse that were harnessed as power sources. Larger animals, such as the elephant and camel, were used where they were available and smaller animals such as the ass, donkey, and dog were frequently used to pull small carts, whims, or to propel treadmills. Figure 1.8 is a detail from a Rowlandson print of 1800. It shows a small dog happily running inside a treadmill and rotating the beef on a spit before the open fire. Such devices must have been quite common at one

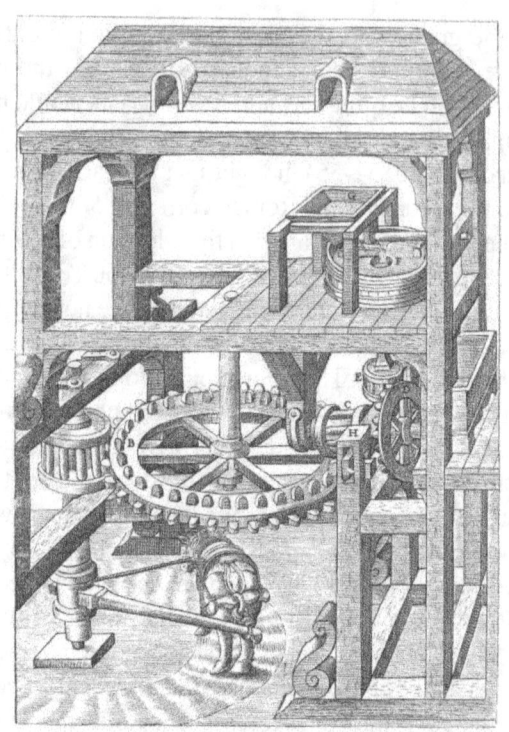

Figure 1.7 Corn mill driven by a horse whim. Ramelli (1588, 311).

Figure 1.8 Dog driving a treadmill in a kitchen. Detail from a watercolour print by Rowlandson, 1800.

time but few have survived. Atkinson (1960, 31) reports that treadmills for dogs survive at the George Inn, Lacock, Wilts, and at the Welsh Folk Museum. In 1750 (July 30th) *The Northampton Mercury* reported a turnspit dog that was mad and was put down after biting several people and on 20th December 1798, *The Derby Mercury* reported a hedgehog at the Angel Inn, Felton, that performed duty in the turnspit in place of the usual dog, see Morsley (1979, 147).

Appendices

The force, work and power that can be developed by animals and humans are often misunderstood. This has not been helped by the fact that engineers, for more than two hundred years, have measured the power output of machines in units of horsepower. This is defined as 33000 ft lb / minute (about 0.746 kW) and has, perhaps, encouraged a belief that the power developed by a horse is invariably one horsepower. In general, we can say that man or beast:

- exerts a maximum force, called the isometric load, when stationary,
- produces maximum power (force times velocity) when the load is about 30%-40% of the isometric load,
- does maximum work and has greatest endurance when the load is about 10%-15% of the isometric load, and
- attains maximum speed when the load is zero.

The metabolic rate is the rate of energy consumption by the body. A commonly used conversion, due originally to Weir (1949, 1-9), is that the metabolic rate in kcal/min is five times the measured oxygen consumption in litre/min (the metabolic rate in Watts is 349 times the oxygen consumption in litre/min). Its value depends on the degree of vigour of the activity undertaken, and typical values and endurance times are listed in Table 1.3. The actual work done depends on the efficiency of conversion, and is typically about 20-25%.

Table 1.3 Metabolic Rates for Various Activities

Activity	Metabolic Rate (W)	Endurance Time
Sleeping	80	Infinity
Standing	125	Infinity
Walking at 1 m/s	230	Infinity
Crawling	410	Finite
Chopping Wood	500	5 hours
Sawing Wood	560	3 hours
Climbing	700	90 minutes
Running at 5m/s	1500	4 minutes

Treadmills

Men or animals working on a treadmill were formerly a common source of power and the best efficiency is about 25% when the climbing rate is about 50 steps/minute. The treadmill was re-introduced into England, and elsewhere, in the nineteenth century to find employment for prisoners, Bennett and Elton (1898, 228). The prison reformer Elizabeth Fry in 1827 thought that extra food should be provided for those put to the treadmill and thought the treadmill useful for "the refractory, the hardened, and the depraved; but- - - ought to be applied to women only under watchful care and with strict limitations". Initially it was thought that an ascent of 12000 feet (about 3600 m) per day was reasonable but this proved to be excessive and was later reduced to 4500 feet (about 1360 m) per day. This is not too arduous for fit young men but it may have been quite demanding for a prisoner in the 19th century. Hill-climbers usually reckon an ascent of 300m per hour is reasonable for the average person and this is equivalent to a work rate of 60 W.

Treadmills are peculiar because the more people treading the mill the greater the work rate demanded from each person. Thus adding an extra person to the mill to share the load has the reverse effect and actually, increases the power demanded of each person. This is because adding an extra person causes the mill to run faster and thus the work rate of all the operatives is increased. For this reason, as mentioned above, it was usual for treadmills used in prisons to be fitted with a speed control governor. Figure 1.9 shows a manually operated a treadmill. The mill rotates with a peripheral velocity, v, and has a radius R. The man's weight is m and local gravity is g, thus he exerts a vertical force mg at his standing position, which is distance r from the centre line of the mill. If there are n people stood in a line at this distance then the torque

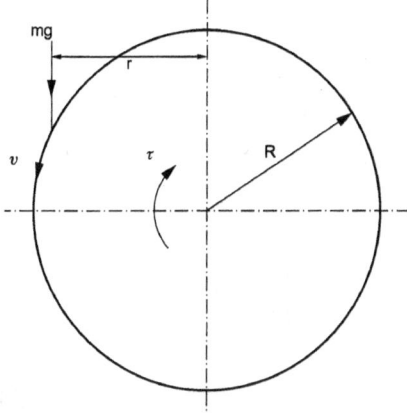

Figure 1.9 Treadmill Geometry and notation.

exerted on the mill is *nmgr*, and the power generated is the product of torque and rotational velocity

$$W = \tau\omega = nmgrv/R \qquad 1.1$$

The power required depends on the nature of the device that the mill is driving but in general it may be determined by assuming the required torque is a constant, τ_0, plus a power function of rotational speed ω_0. This rotational speed is related to the rotational speed of the mill through the overall gear ratio $G = \omega_0/\omega = \omega_0 R/v$. Thus, we assume

$$\tau = \tau_0 + a\omega_0^b = \tau_0 + a(Gv/R)^b \qquad 1.2$$

τ_0, a and b are constants relating to the nature of the load being driven by the mill. The power required to drive the load is then

$$W = \tau\omega_0 = (\tau_0 + a(Gv/R)^b)(Gv/R) \qquad 1.3$$

For a steady state, the power developed by the operatives must be equal to the power required by the load and so from Equations 1.1 and 1.3 we determine that the mill speed is

$$v = \frac{R}{G}\left[\frac{1}{a}\left(\frac{nmgr}{G} - \tau_0\right)\right]^{\frac{1}{b}} \qquad 1.4$$

Substituting this result into Equation 1.1 gives the power developed by each operative as

$$\frac{W}{n} = \frac{mgr}{G}\left[\frac{1}{a}\left(\frac{nmgr}{G} - \tau_0\right)\right]^{\frac{1}{b}} \qquad 1.5$$

Accordingly, as the number of operators increase, the power required of each operator also increases.

To illustrate the power and speed characteristics of a treadmill we may assume $mg = 700$ N, $G = 5$, $a = 2000$ Nms^2, $b = 2$, $\tau_0 = 300$ mN, $r = 1$ m, and $R = 5$ m. The treadmill speed and power per operative for these conditions is shown in Figure 1.10.

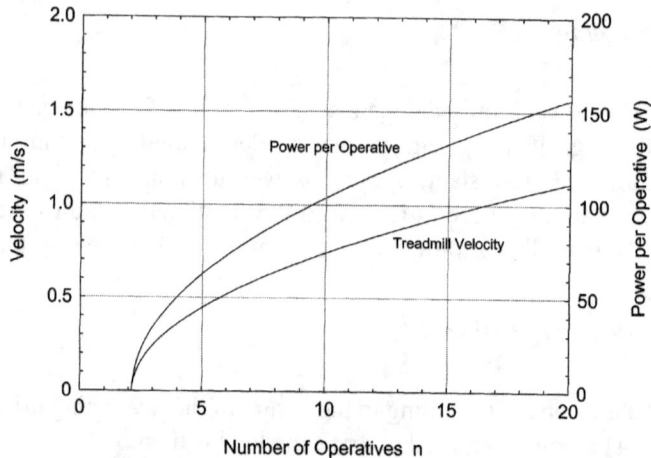

Figure 1.10 Velocity and power demand per person for a human-powered treadmill.

One or two operators standing on the mill are insufficient to start the mill moving but three operators overcome the static torque and accelerate the mill to a steady speed of about 0.25 m/s. As more operators climb onto the mill, its speed increases, following a square root curve. With ten operators, the speed is about 0.74 m/s and with twenty operators, it is a little over 1.1 m/s. All these are reasonable walking speeds but as the operatives are walking uphill the power demand is quite large. For five operators, the power per operator is about 63 W whereas with twenty operators it is about 156 W. This is too high for continuous work by average people so the operators would need to be rested quite frequently.

Muscular Activity in Animals

Brody (1968, 404-469) says that the metabolic rate in mammals at rest is proportional to their body mass, *m*, raised to the power 0.75, that is

$$Q = 3.55 m^{0.75} \qquad 1.6$$

Q is the metabolic rate (W) and *m* is the body mass (kg). Clearly, a 700 kg horse is the equivalent of five or six 70 kg men.

For horses working at light loads, less than 25% of the isometric load, Brody (1968, 901) made some very extensive observations under controlled laboratory conditions. He used the horse treadmill shown in Figure 1.11. Work consists in pulling a weight on a moving horizontal platform actuated by an electric motor

Figure 1.11 Horse treadmill to measure the performance of horses and ponies. After Brody(1968,901).

running at any desired speed and the energy cost was determined from the rate of oxygen consumption. The results were limited to walking speeds because it was considered too dangerous for the horse to trot or gallop on the apparatus.

Numerous ponies and heavy horses were used but the results for horse 19 may be taken as typical. In Figure 1.12 Brody's results have

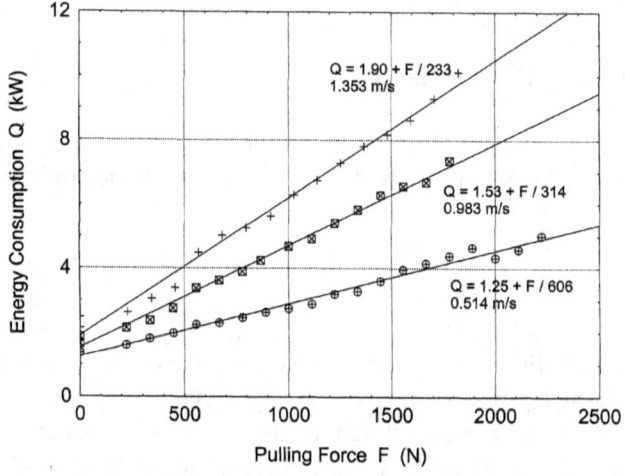

Figure 1.12 Energy consumption rate of a 700 kg horse when pulling various loads at three different speeds. Data of Brody (1968, 953).

been re-plotted and illustrate the linear relation between the energy consumed by the horse and the load it pulled. The intercept on the vertical axis indicates the energy consumed when the horse was walking at the specified speed with no load. Clearly, the energy consumed increases with load and with walking speed. It is more useful to plot the energy consumption, Q, against the power of the horse (the product of load and velocity), as in Figure 1.13.

The lines for the three speeds are now parallel to each other and differ only because the intercept on the vertical axis increases as speed increases. From the definition of the absolute efficiency and gross efficiency we have

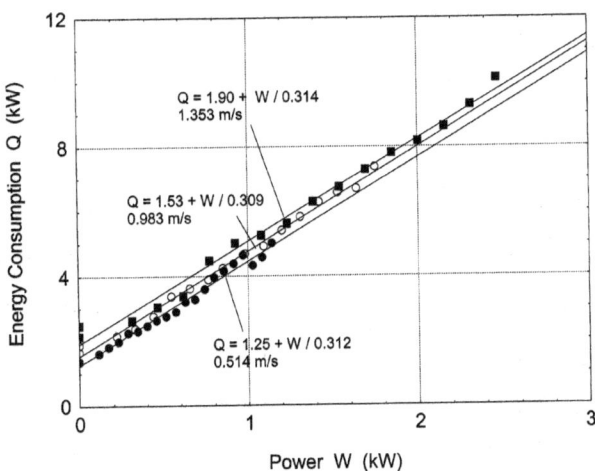

Figure 1.13 Influence of power (force x velocity) on the energy consumption rate of a 700 kg horse. The parallel lines indicate that the absolute efficiency is constant (31%).

$$Q = Q_0 + W/\eta_a$$
$$Q = W/\eta_g$$
    1.7

The intercept on the vertical axis is Q_0 and the slope is the reciprocal of the absolute efficiency. From the figure the absolute efficiency of the horse is virtually constant (31%) and the energy consumption when walking with zero load increases from 1.25 kW at about 0.5 m/s to 1.9 kW at about 1.35 m/s.

The gross efficiency, Equation 1.7, is simply the ratio of power and energy consumption rate. Brody's results show that the gross efficiency rises with increasing load towards the absolute efficiency. At 1 kW the gross efficiency is about 20%. These experimental results may be adapted to estimate the power output of other mammals, such as ponies, oxen, buffalo and elephant, by scaling the results using Equation 1.6.

Chapter 2 Water Power

Introduction

Watermills were the first non-muscular forms of power and were either vertical-axis machines (Norse or Greek mills) or horizontal-axis machines (Roman mills); both used mainly to grind flour. It is not clear which of these was invented first, but as the vertical-axis mill was the simpler to construct (it did not require crown and lantern gearing) it is assumed to be the earliest. They co-existed with the rotary quern and the ancient saddle stone, both of which survived into modern times. In the Middle Ages, other applications for water power were found, for example: to saw logs, to operate bellows supplying air to furnaces, to drive fulling mills, to pump water either to irrigate fields or to drain mines, and to operate stamps to crush the mined ore.

Vertical-Axis (Norse or Greek) Mills

The earliest known mention of a water mill seems to be in a short poem by Antipater of Thessalonica in about 85 BC. Beckman's translation, quoted in Bennett and Elton (1898, 2, 6) is: "Cease your work, you who laboured at the mill. Sleep now and let the birds sing to the ruddy morn. Mother-Earth has commanded the water nymphs to perform your task; and these, obedient to her call, throw themselves on the wheel, force round the axle-tree, and so the heavy mill." There seems little doubt that water was being used to drive the millstones and those who normally rotated the stones by hand (slaves) were, in the view of the poet, free to lie in bed. It is usually assumed that this refers to a vertical-axle mill although, of course, there is no supporting evidence.

Another early reference, of about 60 BC, is to Mithridates' mill and is tantalisingly brief. Strabo (Geography 12.3.30) says "It was at Cabeira that the palace of Mithridates was built, and also a water mill; and here were the zoological gardens, and nearby, the hunting grounds and the mines." Water mills, apparently, were familiar to Strabo's readers and called for no further comment or explanation. Fifteen years later, in about 75 AD, Pliny (Natural History, 28.23), also mentions water mills. "In the greater part of Italy is used a roughened pestle, and wheels also that water turns round as it flows along; and so they mill." Milling was very important to Rome for corn imports formed their staple food. Finley (1965, 35) remarks of the Roman water mill that, "Its use was in one process, corn-grinding, in which there had been a reasonably continuous history of technical advance, a process of immense importance to society, one in which the Roman state, in particular, was immediately concerned. Every 'rational' argument suggests quick and widespread adoption, yet the fact is that, though it was invented in the first century BC it was not until the third century AD that we find evidence of much use, and not until the fifth and sixth of general use. It is a fact that we have no evidence at all of its application to other industries until the very end of the fourth century, and no more than one solitary and possibly suspect reference (Ausonius, *Mosella* 362-4) to a marble-slicing mill near Trier."

Whether these early mills were vertical-axis machines or horizontal-axis machines is not specified, but as mentioned above, vertical-axis machines having a direct drive to the mill wheel were simpler to construct and it is tempting to assume they were the first types to be made. A typical, fully developed, vertical-axis mill is illustrated in Figure 2.1. Water was collected in a millpond by damming a stream and could be released to flow down a chute and impinge with a high velocity onto a paddle wheel. The wastewater then returned to the stream at a lower level. The paddles were usually set at an angle to be roughly normal to the direction of the chute. A pair of millstones was placed over the paddle wheel and a vertical shaft passed through the centre of the fixed lower stone to drive the upper stone. The stones rotated slowly, at a speed less than the water speed, for there was no gearing. Grain was fed into the centre of the top stone from a hopper. Usually the slope of the hopper was adjusted to control the flow of grain. Although not shown in Figure 2.1, the delivery of grain was controlled by the jogging it received from a string between the hopper and a small stone resting on the upper millstone. The rotation of the rough millstone caused the small stone to jostle the string and the hopper. Other devices to joggle the hopper were used. The gap between the upper and lower millstones was usually smaller at the rim than towards the centre so most of the milling was done near the rim. The gap was adjusted to control the quality of the flour by raising or lowering the upper stone by a simple lever arrangement that raised or lowered the bottom cup bearing of the main shaft. Thus, the main shaft, the paddle wheel, and the upper stone were raised or lowered together. Closing a gate in the water chute stopped the mill by diverting the flow into a second chute to by-pass the wheel.

In Northern and Western Europe, this arrangement is known as the Norse mill and it survived in remote places until the last century. Dickinson and Straker (1933, 89-94) measured the performance of a surviving mill at

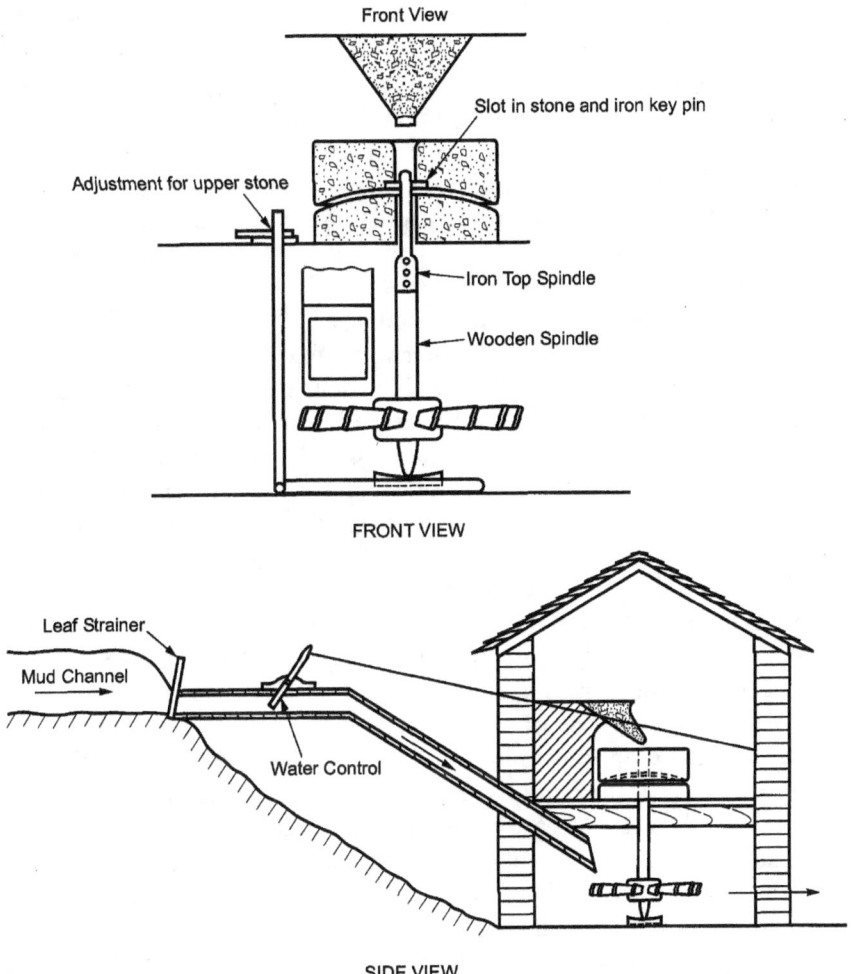

Figure 2.1 Sketch of a vertical-axis water mill or "Ghareth".
After Davidson (1935, 337-369).

Breckon Lock, in the Shetlands. They found that it had seven wooden paddles, perhaps 91 cm diameter, driving an upper millstone of 71-76 cm diameter at a speed of about 50-60 rev/min with a 91-122 cm head of water. It was thought to develop less than 750 W and to have an efficiency of less than 15%. The output of corn was about 0.6-0.75 bushels per hour.

Archaeological evidence for vertical-axis mills has increased greatly in recent times and Rahtz (1981, 1-15) has made an excellent survey.

He mentions vertical-axis machines at Bolle in Denmark, claimed to be of the 1st century AD, and several mills in Ireland firmly dated by dendrochronology to the 7th century, and a mill at Tamworth, England, dated by the same method to the mid-9th century.

The great advantage of the vertical-axis mill was its simplicity, but its slow rotational speed was a handicap without gearing. The maximum torque was generated when the paddles were stationary, but of course, no power was developed. If the paddles moved at the same speed as the water, then the torque was zero and the power developed was again zero. For maximum power, the paddle velocity had to be about half of the water velocity. Thus, if the head of water was 1 m the water velocity was 4.4 m/s and the paddle velocity must be about 2 m/s. For a 1 m diameter paddle this corresponded to a rotational speed of about 40 rev/min, which is usually considered the lowest speed at which flour should be milled. Most vertical-axis mills probably produced less than 750 W but they were easier and more productive than a hand quern. Rahtz (1981, 1-15) says that many hundreds were used in Crete until about 1950-66 when they were replaced by cheap electrical power and internal combustion engines. The head of water had been increased to about 6 m and this allowed milling speeds to increase to about 100-150 rev/min, and power to increase to 1000 -1500 W. The Pelton wheel is the modern equivalent.

An early type of vertical-axis watermill, having a 6-10 m head of water, is the drop-tower mill. Wikander (2000, 376), believes that most of the early, vertical-axis mills are of the drop-tower type and all derive from areas south or south-east of the Mediterranean. Essentially the inclined chute that normally supplies water to the wheel, as in Figure 2.1, is replaced by a strong vertical stone tower filled with water and fitted with horizontal apertures or nozzles in the lower part of the tower walls; these supply a high velocity jet of water onto the wheel. The tower walls need to be quite strong and leak-proof because the water at the bottom is under considerable pressure (6 or 10 m head). The high pressure, low velocity water at the bottom of the tower is converted to a low-pressure, high velocity jet of water as it escapes through the nozzles. For example, a 10 m head of water should give a nozzle velocity of about 14 m/s. If the water wheel rotates at half this speed and is of 1 m diameter, then the rotational speed is 130 revolutions/min. The open inclined chute, on the other hand, does not increase the pressure of the water but converts the potential energy, as it flows down the chute, directly to an increase in kinetic energy. Consequently, the inclined chute can be light and is not subject to the same degree of leakage, but otherwise it has the same performance as the tower. The water velocity delivered by the inclined chute is practically the same as that from a drop-tower of the same head so we must look for other factors to account for the drop-tower's apparent popularity.

Although archaeological evidence for mills in Anglo-Saxon England is growing, documentary evidence is scarce. For example, of the 1300 or so Anglo-Saxon land charters, only a handful mentions a mill. Bennett and Elton (1898, 2, 96-100) found references to a monastic mill in Dover in 762, a mill belonging to Croydon Abbey in 833 and other charters referring to mills in 838, 851, 858, and 963.

In addition to these few references to mills in land charters, a law was passed to prevent the water mills encroaching onto the four principal Roman roads, and mills were subjected to some taxation. This is all. Anglo-Saxon poetry and literature ignores them. It comes as a great surprise, then, to discover that the Domesday Survey of 1086 found, in England, some 5624 water mills in about 3463 settlements, see Hodgen (1939, 261-279). There were, however, 9250 settlements recorded in the survey so about 63% of them had no mill within the settlement. On the other hand, some 506 settlements had several mills, sometimes as many as fifteen. Clearly, mills were placed where there was a suitable stream and this did not always correspond to the distribution of settlements. Large rivers, such as the Thames, Severn and Trent, attracted few mills so they were, evidently, unsuitable for the mills then in use. Their tributaries, however, often had mills little more than a kilometre apart. A few settlements that are now dry were credited with a mill in the survey, so it may be that some horse, oxen, or man-powered mills were in use. The type of water mill, however, is unknown. Some authorities favour the simple vertical-axis Norse mill while others think the horizontal-axis Roman mill was more likely. Archaeological evidence of Saxon mills seems to favour the Norse mill. Most of these were situated south of the rivers Trent and Severn. Some counties, such as Cornwall, possessed scarcely any. The total number of households mentioned in the survey is 287,045 so the average ratio is about one mill for every fifty households, but this varied greatly. In counties, such as Wiltshire, there were 26 households per mill whereas in Devon there were 200 households per mill and in Cornwall 1088 households per mill. Evidently, the humble quern was still used in many parts of England.

The mills varied in rental value from as little as 3d or 4d per annum up to £3 per annum. Some 98% of the Domesday settlements have been located but the precise sites of their mills are unknown. No doubt, many continued in operation on the same site, with much rebuilding, until the eighteenth or nineteenth century. Even the humblest stream could be used, and the site of a water mill may now be found at almost every place where an old road or track leads down to or crosses over a stream. Corn mills were the most frequent, but some paid their rents by weight of metal and are assumed, on this evidence, to be involved in metal or ore processing.

The manorial lord usually owned the corn mill and his corn had precedence. The miller paid rent for the mill and to maintain its value the "soke" laws forced the peasantry to have their corn milled by the miller. The humble quern was made illegal and the miller was empowered to break any he found. It seems likely that the large-scale adoption of manorial water mills and the suppression of the domestic quern owe more to the profit that could be realised from a monopoly than to any concern to reduce the literal daily grind of the peasants. This, of course, led to numerous conflicts. Rahtz (1981, 1-15), for example, notes that that monks at Jumièges, in 1207, broke up large numbers of hand-mills that threatened their monopoly, but the most famous and long-lasting conflict was at St Albans Abbey, which started in 1274 and lasted for over a century.

At some point, the vertical-axis waterwheel declined in favour of the more familiar horizontal-axis machine, but in remote areas, they remained in use even into the 20th century. Out of 65 illustrations of waterwheels in Ramelli (1588), only 14 are vertical-axis machines. In most of these, the wheel is shown submerged in a fast-flowing stream but three designs exhibit a water chute, which directs a jet of water onto the paddles, and is much more efficient.

Another form of vertical axis water mill was the tide mill. Stowers (1957, 239-56) writes that one existed at Dover before 1086 and the Domesday Book says: "at the entry of Dover Harbour, there is a mill, which batters nearly every ship by greatly disturbing the water, and does great harm to the King and to all men". Others were to be found in the first half of the 12th century on the river Adour, near Bayonne in Southern France. Quarr Abbey, a Cistercian Monastery on the coast of the Isle of Wight, operated a tide mill, known as Wootton mill, in the 13th century. The tide regularly filled and emptied the shallow inlet and covered some 25 hectares. The millpond was thus both large and reliable and the mill consequently was very valuable. Such mills were never very common but it is known that there were three in the 13th century, five in the 14th, nine in the 16th, eleven in the 17th, 14 in the 18th and 25 in the 19th century, Gimpel (1993, 23). They continued in use until the 20th century, for example, on

Southampton Water. Keller (1964, 65) illustrates a tide mill proposed by Verantius, Figure 2.2. The vanes of this vertical-axis mill are hinged so that on one side they are forced inward against the frame by the incoming tide and on the other side are forced outwards, and thus it produces a significant torque that rotates the vertical shaft and the geared millstones.

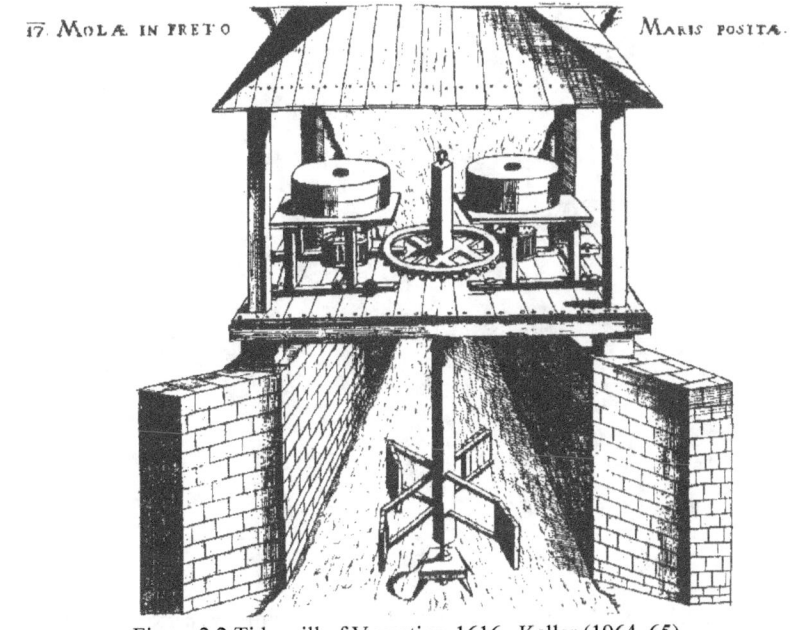

Figure 2.2 Tide mill of Verantius, 1616. Keller (1964, 65).

Branca in 1607 designed the somewhat fanciful vertical-axis waterwheel shown in Figure 2.3 (left). This is really a hydraulic turbine based on a reversed Archimedean screw. A drum C is held on a vertical-axis by bearings, and a pipe is wrapped around the drum in the manner of a screw thread. Water entering at the top accelerates as it flows down the pipe and the reaction caused as it leaves the pipe rotates the drum. It is a form of jet propulsion reminiscent of Hero's turbine. The drum's rotation could be geared to drive the millstones but Branca chooses to complicate the system by arranging it to drive a second and a third drum F and G. It seems unlikely that this design was ever built but a single drum seems quite practical. Knight (1884, 2650) and Bennett and Elton (1898, 2, 28) describe a similar design of vertical-axis wheel from the provinces of Guyenne and Languedoc, Figure 2.3 (right). Here an inverted cone with spiral flutes on its surface was employed to drive machinery through a vertical drive shaft. The cone could rotate in a chamber (not shown) that fitted around the rotor. Water descends a chute at a high velocity and rotates the cone partly by its impulse and partly by its weight on the flutes as it descends to the exit channel.

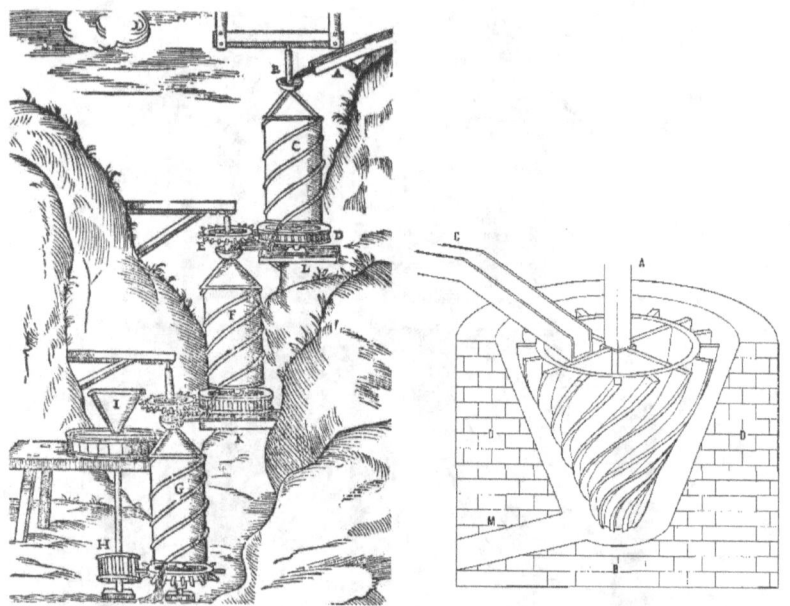

Figure 2.3 Left: Water powered vertical-axis corn-mill. Branca, Le Machine (1607). Right: Vertical-axis conical waterwheel (tub wheel). Knight (1884, 2650).

Horizontal-Axis (Roman) Mills

The problems created by the low speeds and powers of the Norse mill were overcome by the invention of the horizontal-axis, or Roman mill, which was first described by Vitruvius (X, 5) in about 15 BC. He describes paddles fixed to the circumference of a wheel that are driven round by the flow of a river. "Watermills", says Vitruvius, "are turned on the same principle, and are in all respects similar, except that on one end of the axis they are provided with a drum wheel, toothed and framed fast to the said axis; this being placed vertically on the edge turns round with the wheel. Corresponding with the drum-wheel a smaller horizontal toothed wheel is placed, working on an axis whose upper head is in the form of a dovetail, and is inserted into the millstone. Thus, the teeth of the drum-wheel, which is made fast to the axis acting on the teeth of the horizontal wheel, produce the revolution of the mill-stones, and in the engine a suspended hopper supplying them with grain, in the same revolution the flour is produced". There seems little doubt that a horizontal-axis waterwheel is driving a vertical-axis millstone through a crown and lantern gear. Unfortunately, no original illustrations of Vitruvius' water mill survive. His book was not rediscovered in Europe until the 15th century but the earliest reproductions from this period illustrate a horizontal-axis waterwheel driving a vertical-axis mill wheel through crown and lantern gearing.

Parsons (1936, 70-90) discovered the earliest evidence for Vitruvian water mills in the Market Place in Athens close to the Valerian Wall. Using the archaeological evidence and Vitruvius' script he was able to reconstruct the mill, on paper at least, see Figure 2.4. Numerous coins, found beneath the ash layer formed when the mill burned down, suggest that it was active between about 450 AD and 580 AD. Parsons uncovered the millrace, the wheel-pit for a horizontal-axis wheel, the milling room, several millstones, and the bearing sockets for the main shaft. Lime scale on the wall of the wheel pit preserved the imprint of the wheel because the shaft axis was not square with the pit and the wheel tip cut into the lime as the deposit built-up. This deposit, and the positions of the water supply and tailrace, showed that the wheel was overshot; evidently, some progress had been made in the 500 years since Vitruvius. The deposit revealed that the wheel diameter was 3.24 m and it also fixed the centre line of the main shaft. Parsons believes that the water dropped onto the wheel from a height of 1.4 m above the wheel, so here we appear to have a mixed system that relies partly on a jet of high velocity water and partly on the weight of water in the buckets to drive the wheel.

The wooden axle, like the wheel, did not survive the fire that destroyed the mill, but it must have been 3.5 m long and 0.2 m diameter. Its centre was 0.2 m above the floor of the bearing sockets, which leaves space for a set of wooden bearing blocks. Within the mill room, the vertical gear wheel was about 1.11 m diameter, and the diameter of the horizontal gear wheel was 1.36 m, thus the grindstones rotated slower than the water wheel. A shaft passing through the lower stone provided the drive to the upper grindstone. Two complete grindstones, 74 cm and 84 cm diameter, were found together with fragments of four others. They were remarkably thin, the thickest section being only 8 cm. They had shallow notches to receive the drive from the vertical shaft. Some mechanism to alter the gap between the stones was necessary but Parsons found no surviving evidence.

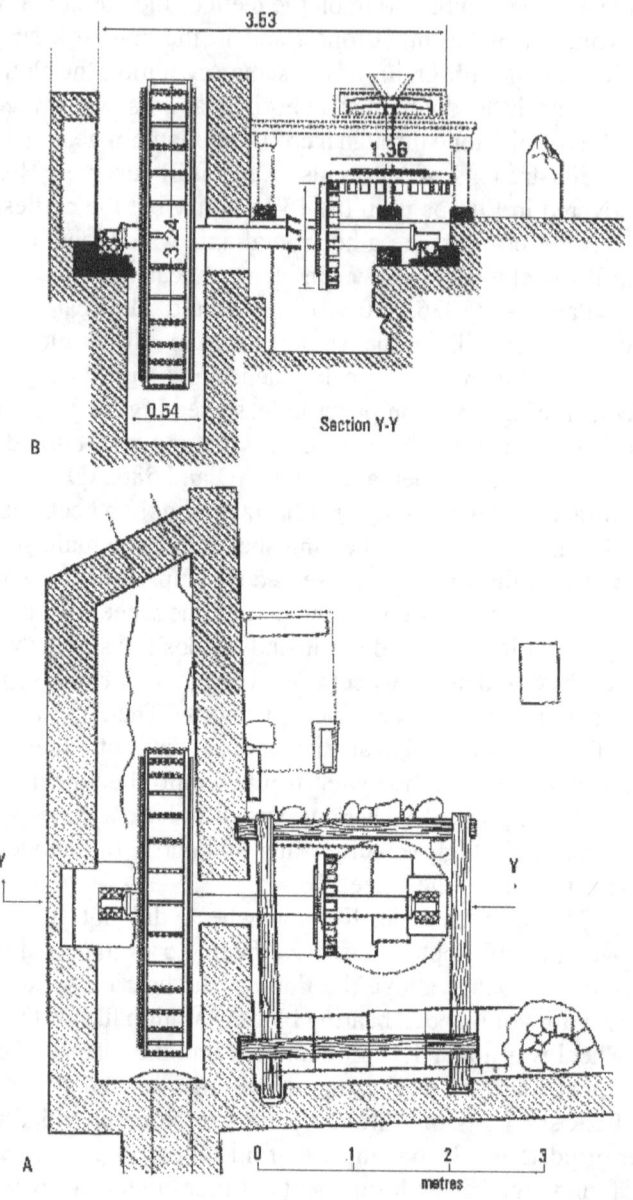

Figure 2.4 Restored plan and section of the horizontal-axis overshot mill in the Athens Agora, after Parsons (1936, 70-90).

However, his reconstruction of the mill, in the light of Vitruvius' description and the archaeological evidence, is very convincing.

With the increasing scarcity of slaves in the Roman Empire, water mills were used more frequently but they were never very common. Nevertheless, by the end of the fourth century, Rome depended greatly on the mills of Janiculum and literary references to mills become more frequent. Brett (1939, 354-356) notes that Palladius recommends the use of the outflow of water from the public baths, which implies that mills were reasonably well known. Prudentius, about the year 390 AD remarks, "What quarter of the city can endure the dire famine, the gradus [the public bread supply] being empty, or what, the motion of the mills of Janiculum being stopped?" This horror was not inflicted until 536 when the Goths besieged Rome and cut the aqueducts supplying the city. To avoid famine the commander, Belisarius, and his engineers, devised a system of floating mills tethered to riverbanks below the bridges where the flow was most rapid. In this way, they could grind sufficient corn to feed the city. Floating mills were occasionally used in later ages. They were installed on the Tigris in the 10th century and in Europe during the 12th century, and they too were often tethered near bridges where the water velocity was high.

The earliest surviving representation of a Roman water mill is a mosaic at the Great Palace at Constantinople, Brett (1939, 354-356). The mosaic is of the fifth century, Figure 2.5, and shows the wheel of the horizontal-axis, undershot water mill. It is not known what the wheel is driving or if there are any gears in the transmission. Another mosaic from the Great Palace of Byzantium, Rahtz (1981, 1-15), apparently depicts the outflow from two vertical-axis water mills, so both types seem to have been in use.

For an undershot wheel, the speed of the wheel depends on the velocity of the stream. The force exerted is greatest when the wheel is stationary and it falls to zero when the wheel rotates with the same velocity as the stream. For maximum power, the wheel velocity at the periphery should be about a third of the stream velocity. This imposes a low rotational speed on the wheel but the necessity of gearing to drive the vertical-axis millstone means that these may be rotated at a much higher speed, although the evidence is that this was not done. The low rotational speed implied a low power output, but very wide water wheels were used in compensation. Undershot water wheels (current wheels) are typically rather wide compared to other water wheels.

Figure 2.5. Early fifth century mosaic showing a Roman water mill. Great Palace, Constantinople. Brett (1939, 35 4-356)

A more impressive system of overshot waterwheels was built during the reign of Constantine at Barbegal, near Arles, France, and consisted of eight pairs of overshot wheels, served by two millraces. They were arranged in series down a 30-degree slope having a fall of 18.6 m. The wheels were 70 cm wide and 220 cm diameter. The millstones were 90 cm diameter and 30 cm deep and each pair is estimated to have a milling capacity of 15 to 20 kg of corn per hour. The whole mill could grind 3.2 tonnes of corn per day and was sufficient to feed the people of Arles, which, in Roman times, had a population of about 8,000.

Water mills were used in Britain during the Roman occupation, although the surviving evidence is scarce, Liversidge (1968, 184). The best example is at Chollerford, near the Roman fort at Chesters. Here, where Hadrian's Wall bridges the river, a stone axle and the remains of a millrace were discovered. A similar millrace and a stone spindle bearing were found further west where the wall crossed the river Irthing. An iron spindle for a waterwheel was found at Great Chesterfield and the remains of two more mills, thought to be sawmills, were discovered on the river Witham at Lincoln. The power output of horizontal-axis water mills was quite high, given that large paddles could be used. At a velocity of 2 m/s, the kinetic energy of water is about 4,000 W/m^2 of which perhaps 1,000 W/m^2 was converted to useful work. It is the indefatigable equivalent of about ten men and lent itself to applications other than corn grinding. Rotational speed was low because for maximum power, as mentioned above, the paddle speed must be about a third of the water speed. However, the necessity for crown and lantern gearing to drive the vertical-axis millstones enabled a suitable gear ratio to be chosen so the millstones turned at their optimum speed (usually between 50 and 300 rev/min).

One of the first industrial uses of waterpower was in the manufacture of wool. Fulling mills were used in Italy during the tenth century, and elsewhere in the eleventh. In England water mills for fulling cloth were not mentioned in the Domesday Survey of 1086 but were used in the latter half of the twelfth century, Carus-Wilson (1941, 39-60). The earliest English fulling mill dates from 1185 and was built on land of the Templars at Newsham, Yorkshire, and another mill of the same date was built at Barton in the Cotswolds. Quarr Abbey on the Isle of Wight had a fulling mill in 1189, Hockey (1970, 50). The need for fulling is given by William Langland, in the fourteenth century, who wrote "Cloth that cometh from the weaving is naught comely to wear, till it is fulled underfoot, or in the fulling stock". Despite the early invention of the fulling mill, treading cloth underfoot continued in the more remote parts of the British Isles until comparatively recent times, as Figure 2.6 (top) illustrates. The texture of freshly woven cloth from the loom must be cleansed of any grease remaining in the wool and of the size applied to strengthen the warp during weaving. Fuller's earth or soft soap was used, and pummelling or beating the cloth in water cleaned and thickened it. After fulling, the wool cloth was rinsed and dried,

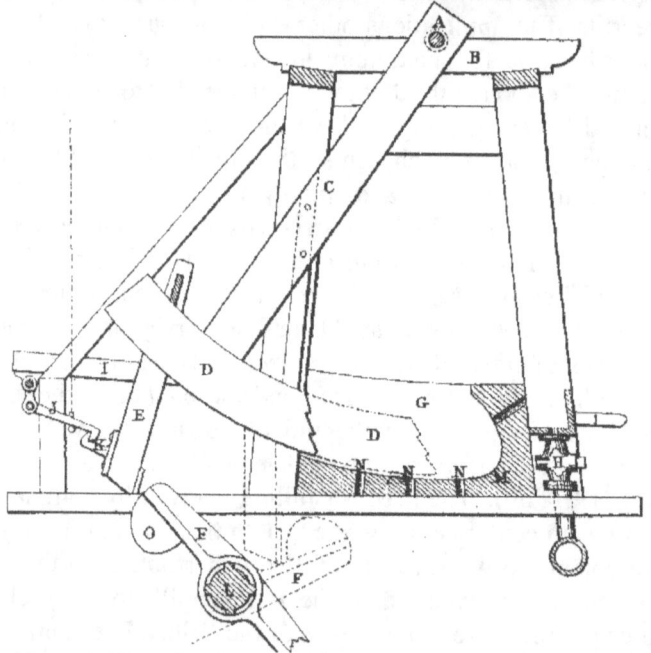

Figure 2.6 Top: Treading cloth in Connemara, early 20th century. Bottom: a fulling mill. McCutcheon (1967, 67-94).

shrinking markedly in the process to become thicker, softer and dimensionally stable.

Men or women trampling on the cloth in a trough were released from their toil by the invention of large wooden hammers operated by a rotating cam and driven from a waterwheel. Figure 2.6 (bottom) shows such a fulling mill. McCutcheon (1967, 67-94). As the shaft L rotates the cam F and follower E raise the heavy wooden stocks D and release them so they fall back into the washing chamber G. The wooden stocks weigh about 250 kg and they compress the cloth in the chamber squeezing out the water and the impurities that drain away through the holes N. The water is replenished through valve H. A lever J allows the stocks to be held when required. Such devices were not always welcomed for they were a threat to the livelihood of traditional fullers. In 1298, their use in London was forbidden and occasional opposition continued for centuries.

Nevertheless, fulling was neither an easy nor a pleasant task so it is hardly surprising that fulling-stocks became popular. The clear majority of the earliest fulling mills were on Abbey lands or Episcopal estates although some were built on lay holdings. By the beginning of the 14th century, fulling stocks had mostly replaced the more primitive method of treading underfoot. Carus-Wilson (1941, 39-60) points out that a profound change in English cloth manufacture was underway. The new mills were usually situated in hilly districts having a plentiful and rapid water supply. The cities of York, Beverley, Lincoln, Louth, Stamford and Northampton were once famous for their fine quality cloth, but, lying on flat land, they had slow moving rivers on which undershot waterwheels (current wheels) were necessary and power output was too low to drive fulling mills. Consequently, their industry declined and then vanished. Cloth manufacture moved out of towns and cities to the hilly countryside where rapid streams and steep valleys enabled overshot wheels of greater power and torque to be used. By the 14th century there were 30 fulling mills in the West Riding of Yorkshire and the Lake District, and 34 in the West of England and the Cotswolds and very few in the low-lying areas of eastern England.

Water mills, by the 16th century, were being applied to a variety of operations. They were used: to grind corn, to lift the stocks of fulling mills, to drive water pumps, to raise water and ore from mines, to operate trip hammers in the forge, to operate bellows providing air to furnaces, and to lift stamps to crush ore. A good practical understanding of water mills had developed. Stowers (1957, 239-56) quotes a short but interesting comment on the design of watermills written by John FitzHerbert and published posthumously in 1539: "there be two manner of corn mills, that is to say a breast mill and overshot mill, and these two manner of mills be set and go most commonly upon small brooks and upon great pools or meres and they have always a broad bow a foot [30 cm] broad or more, and the ladles be always shrouded with compost boards on both sides to hold in the water, and then they be called buckets. And they must be set much closer together than the ladles be, and much more a slope downwards to hold much water that it falls not out for it driveth the wheel as well with the weight of water as with the strength. And the miller must draw his water according to his buckets, that they may always be full and no more, for the longer that they hold the water the better they be."

The principle enunciated here is very sound and is best illustrated in Figure 2.7. A simple overshot wheel with radial paddles is shown in Figure 2.7 (left). Even if the paddles are fitted with shrouds to prevent water escaping from the sides, the water nevertheless escapes over the outer edge of the paddle as it rotates. By "sloping the paddles downwards" as

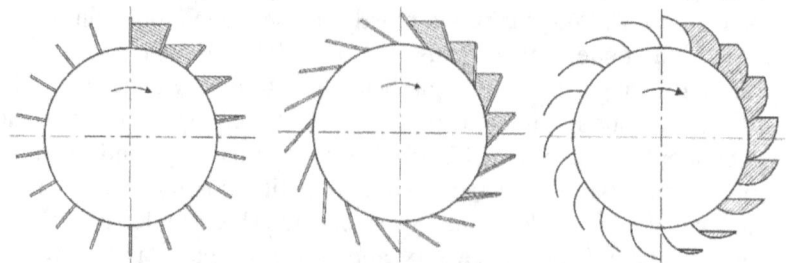

Figure 2.7 Improvements in the water retaining capacity of waterwheels.
Left: radial paddles. Centre: sloping paddles. Right: curved paddle.

Fitzherbert recommends, Figure 2.7 (centre) a much greater proportion of the water is retained in the paddle and this increases the torque, power and efficiency of the wheel. Figure 2.6 (right) illustrates the use of curved paddles. These improve the torque, power and efficiency even further but they are difficult to make in wood and were not adopted until the 19th century when metal waterwheels became common.

Fitzherbert's ideas must have been reasonably well-established practice, for a few years later, in 1588, Ramelli published a drawing of an overshot waterwheel with shrouds and inclined paddles, Figure 2.8. Although the waterwheel is particularly interesting, the application is unfortunate. The illustration shows a kind of perpetual motion machine in which the waterwheel not only drives a corn mill but also drives a pump

Figure 2.8 Overshot waterwheel with shrouds and sloping paddles. The wheel drives a corn mill and a pump is used to lift the wastewater back to the millpond. It is a form of perpetual motion machine. Ramelli (1588, 297)

that lifts the wastewater back to the millpond so it can be re-used. Ramelli says: "[The mill] is helped by having part of the water which falls over the wheels return to the pool or fountain after it has produced its effect." He is describing, of course, a perpetual motion machine and we now know that friction and other losses will cause it to stop. Even practical and experienced

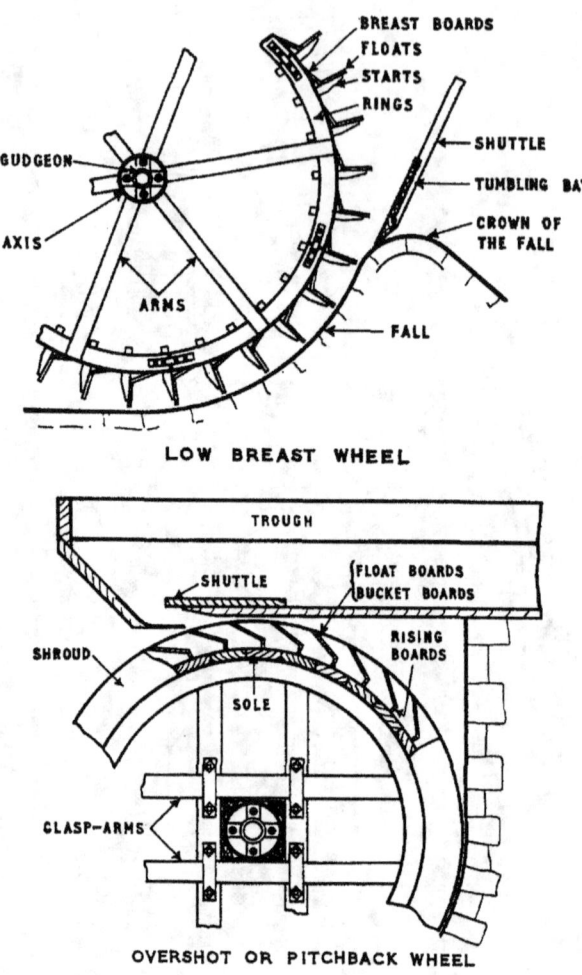

Figure 2.9 Typical low breast and overshot waterwheels developed by Smeaton. Wilson (1955, 37-38).

engineers, such as Ramelli, were capable of such blunders before a proper understanding of energy evolved.

John Smeaton (1724-92) was the first to investigate the performance of waterwheels in a rigorous manner, see Wilson (1955, 37-8), and in his models, Smeaton obtained maximum efficiencies of about 22% with an undershot wheel and about 63% with an overshot wheel; results that he communicated to the Royal Society in 1759. According to Wilson,

Smeaton's wheels marked the conclusion of eighteen centuries of development for wooden waterwheels; they had reached the limit of their power and the future lay with his introduction of cast-iron parts and ultimately with all-metal construction. The breast wheel, Figure 2.9 had a stone wall to prevent water escaping from the paddles and so increase the torque, and Smeaton's overshot wheel also used this idea to retain water in the paddles for virtually the whole of its descent, consequently both wheels were very efficient.

Despite the development of steam power in the 18th century, waterwheels continued in use, and were improved, until well into the 19th century. In 1800, in Lancashire, Yorkshire and Cheshire, waterwheels developed by far the greater proportion of motive power and there were only 125 steam engines. Thirty-five years later there were still 866 waterwheels but the number of steam engines had risen to 1,369, Stowers (1957, 239-56). Water wheels continued to develop (all-metal construction, curved paddles, vented buckets) but their end was nigh for they could not compete with the power and speed of steam engines.

Appendices

Power Output of Water Wheels

The total head, z_0, of a stream of water in the head-race, relative to the tail-race, is

$$z_0 = z + v^2/(2g) \qquad 2.1$$

z is the difference in height between the head-race and the tail-race and v is the water velocity in the head-race. Thus, the hydraulic power, W_h, supplied to the waterwheel is

$$W_h = mgz_0 \qquad 2.2$$

m is the mass flow rate of water, g is local gravity, ρ is the density of water, and z_0 is the total head. The power output of a waterwheel is usually expressed as some fraction of the hydraulic power, that is

$$W = \eta W_h = \eta mgz_0 = \eta mg\left(z + \frac{v^2}{2g}\right) \qquad 2.3$$

η is the efficiency of the wheel. This varies depending on the type of waterwheel. Rankine (1889, 269), states that for overshot and breast wheels that rely mainly on the weight of the water in the buckets to drive the wheel, the efficiency of well-designed wheels may be up to 70-80%. Undershot wheels, which rely mainly on the impulse from a high velocity stream, are much less efficient but may be as high as 40-60%. If the wheel is submerged, then the efficiency is reduced to about 75% of the same wheel when not submerged. Earlier waterwheels, of course, were less efficient than those quoted by Rankine. Smeaton, as mentioned above, measured efficiencies of 22% for undershot wheels and 63% for overshot wheels.

The efficiencies quoted above are the maximum efficiencies that occur when the wheel tip velocity is at its best speed relative to the water flow. For an impulse wheel this optimum speed is about 50% of the water velocity but for current wheels it is about 30% of the water velocity. Further analysis of the mechanics of water mills, set out in the following sections, enables the efficiency, and the factors affecting efficiency, to be determined. The driving force is generated by three methods. For the horizontal-axis, undershot watermill, often called the Roman mill, the driving force is generated from the drag of a paddle inserted into a fast-flowing stream of water. For the horizontal-axis, overshot water mill and the breast mill, the driving force is obtained from the weight of water in the buckets. For the vertical-axis, water mill, known as the Norse mill or the Greek mill, and for the Pelton wheel, the impulse and change in direction of a high velocity jet of water provides the driving force. Often, Greek and Norse mills had paddlewheels that were submerged in the water, which detracted from their efficiency, but in the better designs the paddle rotated above the tailrace so that when not being propelled by a jet of water, they rotated in air.

Horizontal-Axis Undershot Wheels (Roman Mills)

Figure 2.10 (top) shows a diagram and notation for a horizontal-axis undershot waterwheel such as might be inserted into a fast-flowing river or stream. Water flows from the right at velocity v and the wheel rotates with angular velocity ω. The wheel width is w, its maximum depth

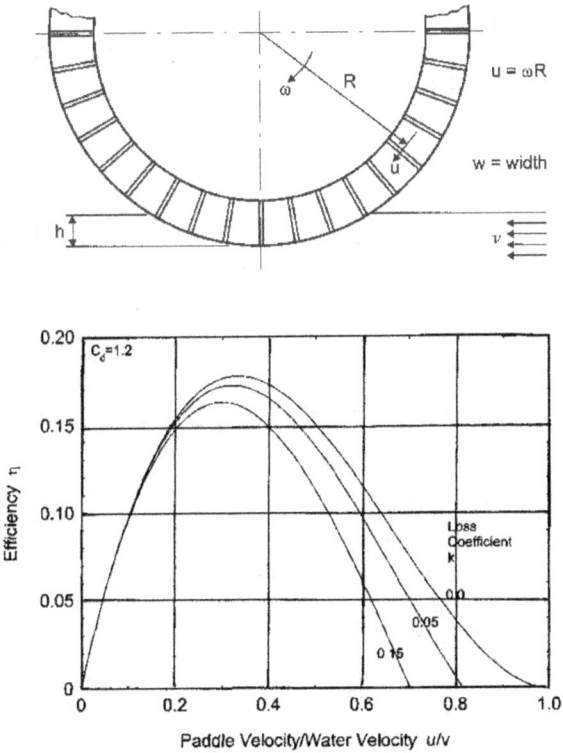

Figure 2.10 Top: diagram and notation for a horizontal-axis waterwheel immersed in a stream. Bottom variation of power output with rotational speed for a horizontal-axis waterwheel, showing the influence of windage and frictional losses.

of immersion into the stream is h and the radius at the average depth of immersion is R. The paddle velocity is thus $u = \omega R$. Assume that the projected area of all the paddles at any instant is wh and the drag coefficient is C_d. The torque, τ generated by fluid drag over the projected area of the paddle wheel is the product of drag force and radius, thus

$$\tau = C_d \rho w h \frac{(v-u)^2}{2} R \qquad 2.4$$

C_d is the drag coefficient, ρ is the density of water, w is the paddle width, h is the depth of immersion, u is the paddle velocity, v is the water velocity, and R is the mean radius of the wheel. The net power generated by the wheel is the product of torque and rotational speed less the power lost to windage and friction

$$W = C_d \rho w h u (v-u)^2 / 2 - k m u^2 / 2 \qquad 2.5$$

k is the loss coefficient, and m is the mass flow rate of water. The mass flow rate of water supplied to the immersed paddles is

$$m = \rho w h v \qquad 2.6$$

From Equation 2.5 and 2.6 the net power may then be expressed as

$$W = \frac{mv^2}{2}\left[C_d \frac{u}{v}\left(1-\frac{u}{v}\right)^2 - k\left(\frac{u}{v}\right)^2\right] \qquad 2.7$$

The efficiency of the waterwheel is the ratio of net power and the rate of kinetic energy supplied by the water, thus

$$\eta = \frac{2W}{mv^2} = C_d \frac{u}{v}\left(1-\frac{u}{v}\right)^2 - k\left(\frac{u}{v}\right)^2 \qquad 2.8$$

This relation is plotted in Figure 2.10 (bottom) where, for the ideal waterwheel, the power output is a maximum when the paddle speed is a third of the speed of the stream. This leads to a relatively low rotational speed but this is not necessarily a problem because it may be geared to drive a set of grindstones having a vertical-axis. The gear ratio may be chosen to suit the speed of the stream and the required speed of the stones. By differentiating Equation 2.8 with respect to u and equating it to zero, it is easily shown that the conditions at maximum power are, approximately,

$$\eta_{max} = \frac{4C_d}{27} - \frac{k}{9}$$

$$\frac{u}{v} = \frac{1}{3}\left(1 - \frac{k}{C_d}\right)$$

2.9

The maximum power developed by a horizontal-axis water mill is thus

$$W_{max} = \left(\frac{4C_d}{27} - \frac{k}{9}\right)\frac{\rho w h v^3}{2}$$

2.10

This occurs when the paddle speed is about 30% of the water velocity and the efficiency is about 16%.

Example: A Horizontal-axis Undershot Waterwheel

Determine the paddle speed and rotational speed of a 2 m diameter horizontal-axis undershot waterwheel if the projected area of immersion (wh) is 1 m². The speed of the river is 3 m/s, the drag coefficient of the paddles is 1.2, and the loss coefficient is 0.1. What is the power output and torque under these conditions if the density of water is 1000 kg/m³.

From Equation 2.9 η_{max} = 16.7% and when the speed ratio is u/v=0.306. Thus, the paddle speed is 0.917 m/s and the wheel speed is 8.75 rev/min. From Equation 2.10 W_{max} = 2.25 kW and the torque is 2455 Nm.

Example: A Horizontal-axis Overshot Waterwheel

The paddles of an overshot waterwheel have a mean diameter of 2 m and are supplied with water in a chute 2 m wide and 0.5 m deep. The water velocity is 3 m/s. If the efficiency is 60% determine the output power and rotational speed. The density of water is 1000 kg/m³.

The mass flow rate is 1000x2x0.5x3=3000 kg/s.
The total head (Equation 2.1) is 2+3²/2/9.81=2.46 m
The power output (Equation 2.3) is .6x3000x9.81x2.46=43.4 kW
The rotational speed is v/πD=3/3.142/2=0.477 rev/s=28.6 rev/min.

In this example, the overshot wheel is supplied with the same volumetric flow rate of water as the undershot wheel in the previous example. The overshot wheel produces 16 times more power. This is because the static head of the supply has increased from 0.46 m to 2.46 m, the efficiency has increased from 17% to 60%, and the rotational speed has increased from 8.75 rev/min to 28.6 rev/min. It should be noted that the useful power output of both these machines would be reduced by the transmission losses associated with the gears that would normally be attached to the output shaft. A transmission efficiency of 60-70% may be reasonable.

Vertical-Axis Wheels (Greek or Norse Mills) and Pelton's Wheel.

Vertical-axis machines, such as the ancient Norse or Greek mill, the drop-tower mill, and the relatively modern Pelton wheel, which has a horizontal-axis, may be treated by the same analytical methods. This is because they all rely on the impulse and change in direction of a high-velocity jet of water impinging against a paddle on a rotating wheel.

Figure 2.11 shows the general arrangement and notation for horizontal axis water wheels. A high velocity jet of water at velocity v and mass flow rate m impinges on a paddle moving at velocity u. The paddle may be carefully streamlined to turn the water jet so it leaves at angle θ, although the earliest designs used flat paddles (θ=90 deg). The water velocity, as it travels over the paddle's surface loses velocity by friction and leaves at relative velocity of k(v-u) (k is approximately 0.9). The force exerted by the water on the paddle is the product of the mass flow rate from the jet and the change in velocity and the direction of motion, thus

$$F = m[(v-u)+k(v-u)\cos\theta]$$
$$F = m(v-u)(1+k\cos\theta)$$

2.11

The net power output is the product of this force and the paddle velocity less the power dissipated by the drag of the paddle.

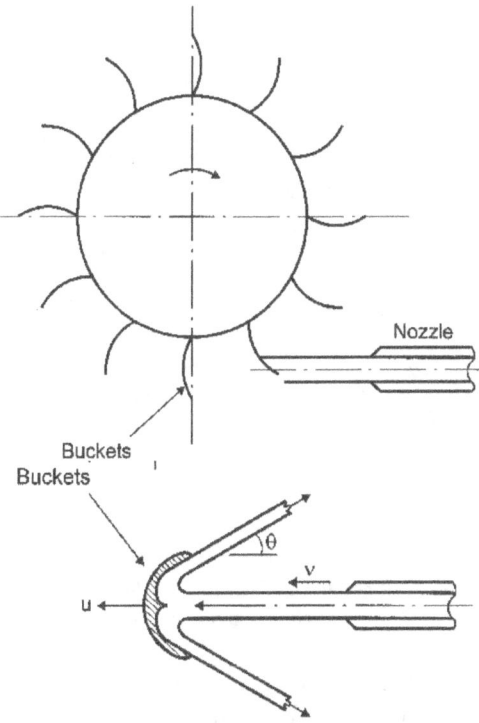

Figure 2.11 General arrangement and notation for horizontal axis water wheels.

$$W = Fu - mC_d u^2 / 2$$

$$W = \frac{mv^2}{2}\left[2(1+k\cos\theta)(1-u/v)u/v - C_d(u/v)^2\right] \quad 2.12$$

C_d is the drag coefficient. The efficiency of the system is defined as the ratio of the net power output to the rate of energy supplied in the jet, thus

$$\eta = \frac{2W}{mv^2} = 2(1+k\cos\theta)(1-u/v)u/v - C_d(u/v)^2 \quad 2.13$$

Equation 2.13 is plotted in Figure 2.12 for the ideal case in which the jet is completely reversed ($\theta=0°$) and there are no losses ($k=1$ and $C_d=0$), for the more practical modern case in which $\theta=20°$, $k=0.9$ and $C_d=0.8$, and for the original mills when $\theta=90°$, $k=0.9$ and $C_d=0.8$.

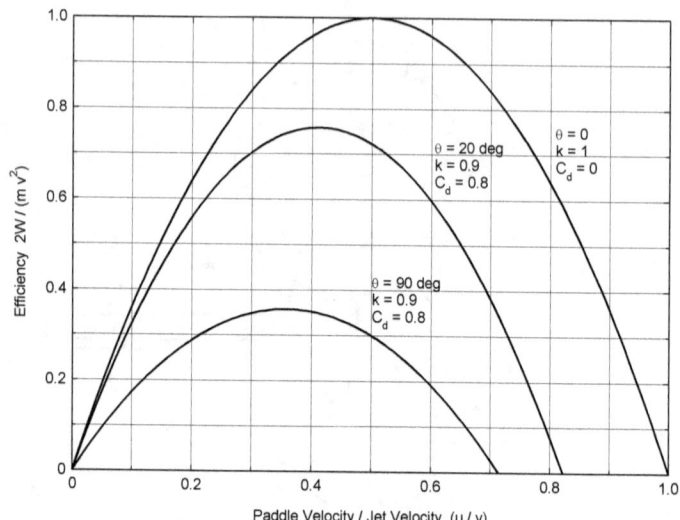

Figure 2.12 Efficiency of an ideal Pelton wheel and horizontal axis watermills wheel with frictional and windage losses.

In the ideal case the maximum efficiency is 100% and occurs when the paddle velocity is half the water velocity. In the second case the maximum efficiency is about 76% when the paddle velocity is about 41% of the water velocity, and in the original mills ($\theta=90°$) the maximum efficiency is 37% when the paddle velocity is 31% of the water velocity.

The speed ratio at which efficiency is a maximum is determined by differentiating Equation 2.13 with respect to u/v and equating to zero. This condition yields

$$\left(\frac{u}{v}\right)_{max} = \frac{1+k\cos\theta}{2(1+k\cos\theta)+C_d} \qquad 2.14$$

The maximum speed the paddle can attain is exactly twice this value and is the speed at which efficiency is zero. If the load is removed from the wheel then it accelerates to this maximum speed and it is, therefore, referred to as the runaway speed. Typically, it is about 80% of the water speed from the jet.

> **Example: A Vertical-axis (Norse) Waterwheel**
> A Norse mill has a mean diameter of 2 m and is supplied by water in a chute 0.25 m deep and 0.25 m wide. A head of 2 m generates the water velocity. If the water leaves the rotor at $\theta=90°$ determine the best rotational speed and the efficiency and power output at this speed. The loss coefficient C_d is 0.7. and k=0.1. Density of water is 1000 kg/m^3.
>
> The water velocity for a head of 2 m is $(2 \times 9.81 \times 2)^{0.5} = 6.26$ m/s
> From Equation 2.14 $(u/v)_{max} = 0.3704$. Thus u = 2.32 m/s
> Rotational speed is $N = u/\pi D = 0.369$ rev/s = 22.2 rev/min.
> Mass flow rate is = 391.25 kg/s
> From Equation 2.13 efficiency is 0.3696 = 37%
> From Equation 2.12 the power is 2836 W = 2.84 kW.

The Norse mill rotates 2.5 times faster than the Roman undershot mill on page 45, and it is more efficient. It produces only a little more power than the Roman undershot mill because the chute area and volume of water flow is much smaller. It is not surprising, therefore, that the Norse (or Greek) mill remained popular where water supplies were scarce.

Chapter 3   Wind Power

Introduction

It is astonishing that windmills were not developed until about 3,500 years after the first use of sails. By 2700 BC, the primitive reed boat of the Nile had developed into a more substantial structure capable of transporting rock and stone from quarries in the south to building sites in the north. From this time on there is a wealth of illustrations depicting river craft and sea going vessels with sails. Sailing ships show a continuous development from 2700 BC onwards and were very common, but the application of wind power to other devices did not develop. Apparently, the Greek and Roman civilisations knew nothing of the use of wind in generating power, although Hero of Alexandria describes a wind organ "with oar-like scoops like the so-called wind-motors". But nothing more is known of these wind-motors.

Vertical-axis Windmills

Forbes[2] (1955, 114) says that the earliest windmill emerges during the reign of the Caliph Umar ibn al-Khattab (634-44 AD). A captured Persian named Abu Lu'lu'a claimed under interrogation to be able to build a mill that could be rotated by the wind. The Caliph was intrigued and commanded him to build one. This ultimately proved to be a dreadful mistake, for Abu Lu'lu'a, after building his mill and becoming wealthy, assassinated the Caliph, apparently in protest against high taxes. In about 950 AD we hear of windmills in the Persian province of Seistan, on the borders near Afghanistan, where "strong winds prevail, so that, in view of these, mills were built rotated by the wind". There are several later references to windmills in Seistan, which make it clear that the windmills were used not only for grinding corn but also to pump water from deep wells to irrigate the land. Al-Dimashqi (1256-1327) says that in Seistan they build their windmills on the tops of minarets, hills, high mountains, or the towers of castles. The mills are of two storeys. The lower story contains grindstones driven by a vertical-axis windmill in the upper storey. Four tapered holes in the walls, one for each quarter of the wind, direct the wind onto the paddles of the windmills. The holes are offset from the axis of the drive shaft and the mill is started or stopped by opening or closing the appropriate holes.

Wailes (1968, 125-145) has made an excellent survey of vertical-axis windmills, including many of the surviving mills of Seistan. Figure 3.1 (left), designed by Besson in 1578 is the earliest known illustration of a vertical-axis windmill. They have the great convenience that the sails

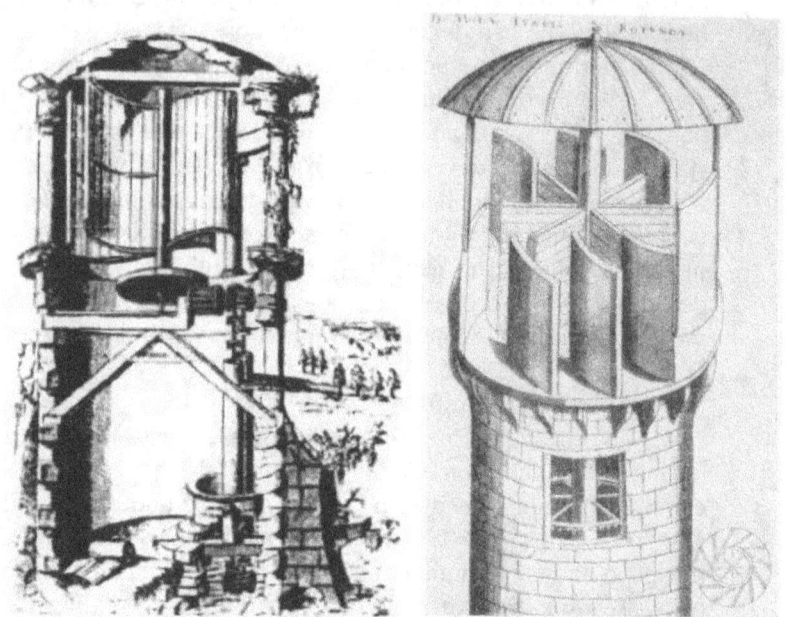

Figure 3.1 Left: a vertical-axis windmill illustrated by Besson, 1578, operating a rag and chain pump for raising water. Right: shrouded wind-wheel of Veranzio, 1595.

do not have to be turned to face the wind, as on a horizontal-axis windmill, but a disadvantage is their low power output and slow rotational speed. The early windmills of Seistan probably had direct drives to the stones and consequently were slow. Besson's windmill is geared so the millstones rotate faster but the power output remains low. The kinetic energy in a wind of 10 m/s is about 600 W/m^2 so the power output from a wind driven mill is unlikely to exceed about 100 W/m^2. Thus, the small area of the windows limited the power input and the vertical-axis windmill could not compete with the more common horizontal-axis machine, the area of whose sails was perhaps a hundred times larger. Vertical-axis windmills continued to be designed, patented, and occasionally built, but they were never very common.

Veranzio (1595) illustrates five designs of vertical-axis windmill. The design shown in Figure 3.1 (right) consists of a vertical-axis rotor having eight radial vanes and set on top of a tower. The wind is directed onto the vanes by a series of deflectors or shrouds so that from whatever direction the wind blows it is deflected to strike the vanes causing them to rotate and to drive the machinery in the room below. It is a remarkable example of an early turbine, as the figure makes plain. However, as drawn, it is unlikely to produce sufficient power, except in a high wind, and to work well the area of the vanes needs to be increased.

Figure 3.2 (top) shows another of Veranzio's designs. This has four hinged sails covered with cloth and mounted above a tower containing two sets of millstones. It is like the tide mill of Verantius, 1616, shown in Figure 2.2 of Chapter 2. The hinged shutter continued into the 20th century as Figure 3.2 (bottom) confirms. In this case, the hinged shutters are not very wide but are quite deep and are mounted at a large radius to give adequate torque. Wailes (1968, 125-145) records many examples of vertical-axis windmills and although they were uncommon they were never abandoned and were reinvented or rediscovered from time to time. Many modern wind power generators are vertical-axis machines.

Horizontal-axis Windmills

Exactly how the windmill emerged in the west is something of a mystery. They first appear in the twelfth century and it is sometimes said that they were based on eastern windmills seen during the Third Crusade. However, the western windmill, which was a horizontal-axis machine, was fundamentally different from its eastern forebear, which was a vertical-axis machine. Essentially the western windmill has four vertical sails on a horizontal wind-shaft, which must be turned to face the wind as it changed quarter, and it drives vertical-axis millstones through crown and lantern gearing, Figure 3.3. It owes more to the construction of the horizontal-axis water mill, which, as we have seen, was very common by the twelfth century. It appears likely to have been invented independently, perhaps even in England where many of the earliest references are found.

The first reliable records of windmills in Western Europe are from the last half of the 12th century. Hills (1994, 37) quotes an entry from the *Testa de Nevill*, that refers to a windmill at Ilford in 1155,

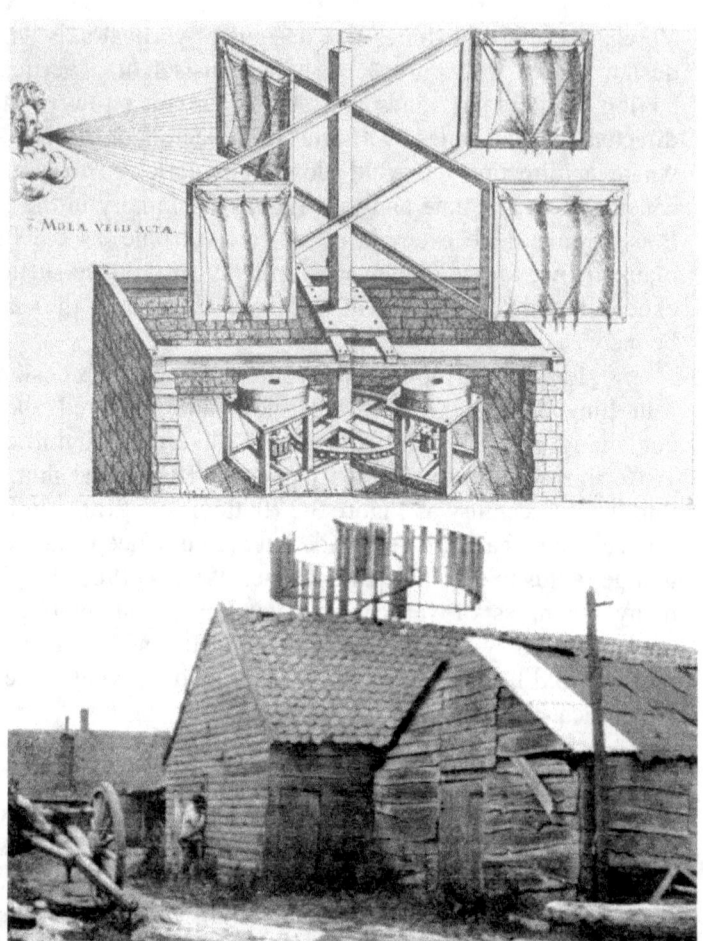

Figure 3. 2 Top: a vertical-axis windmill proposed by Veranzio, 1595, driving a pair of millstones. Bottom: articulated wind-wheel driving machinery in a carpenter's shop at Kieken-Put, Wormheudt, Nord, in 1938.

namely: "Hugo de Plaz gave to the monks of Lewes, the windmill in his manor of Ilford, for the health of the soul of his father". Other early references are quoted by Hills and he finds that "a further twenty windmills that can be dated confidently to the 1190s". It appears that windmills emerged sometime before 1155 and spread rapidly, sometimes causing considerable disruption to existing arrangements.

Figure 3.3 Top: an early English post-mill of the 14th century. MS Bodley 264, Keil, I. Bottom: engraved memorial from Lynn, Norfolk, showing and early 14th century post mill.

Windmills had the advantage that they could be built almost anywhere that the wind blew with sufficient strength. By the middle of the 13th century, there are references to windmills throughout England. They were built, mostly, in manors where, for want of water, it had not been possible to build water mills, but this was not always the case. At Highworth, Wiltshire, for example, a windmill replaced an existing water mill on the lands of Adam de Stratton, one of the king's chief financial advisors. Presumably, it was cheaper to build a windmill than to repair the old water mill. The newly built windmill in 1285 cost £12 2s 10d (£12.14). In 1289, when Adam was disgraced and imprisoned, his lands reverted to the king and the old water mill was found in a ruined condition. New millstones for the watermill and the windmill were purchased from Tewkesbury and carried to Highworth at a cost of 37s (£1.85), see Pugh (1970).

The earliest windmills were based on the post mill arrangement. Many illustrations appear in illuminated manuscripts during the fourteenth century and at Lynn in Norfolk a post mill is illustrated on the memorial brass of Adam de Walsokne, who died in 1349, Figure 3.3. The story depicted on this brass is now lost but it seems to have illustrated a well-known medieval joke; a man on the horse is shown carrying the sack of corn on his shoulders to avoid tiring the horse! The scene is repeated elsewhere Van Belle (2005, 222).

The first drawing showing the interior construction of a windmill is by Ramelli (1588, 335), Figure 3.4 (left). This too is a post mill and seems to be of a type in common use for up to four hundred years. The four sails turn a horizontal wind-shaft that drives a vertical main-shaft through crown and lantern gears. The whole of the mill may be turned, about a fixed post, into or out of the wind as required, by means of the rudder marked S. Turning out of the wind to stop the sails is quite inconvenient and a feature of Ramelli's design is a brake ring that circles the toothed wheel on wind-shaft. This ring may be

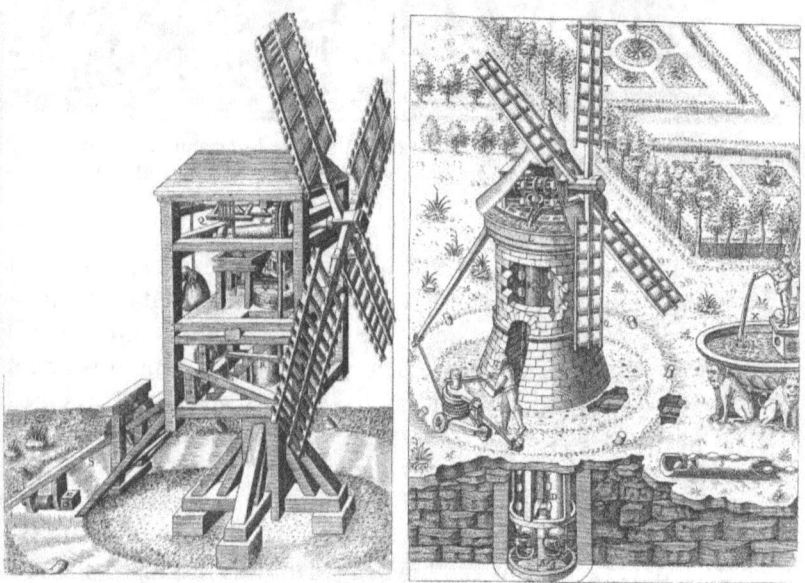

Figure 3.4 Left: a post mill driving millstones, Ramelli (1588, 335). Right: a tower mill driving a pump to lift water from a well for irrigation. Ramelli (1588,206).

tightened or loosened with bar R that is connected to it by a rope. Thus, the miller has complete control of the sails and of the grinding process and he may rotate the mill to take full advantage of the wind or to take the sails out of the wind during storms. Sacks of grain for milling are lifted to the top floor by a rope wrapped around drum Q and turned from the second floor by another rope. In a later development, the sack was raised by a rope-drive taken from the wind-shaft.

Another windmill illustrated by Ramelli is the turret mill shown in Figure 3.4 (right). Here, only the cap, the sails and the rag and chain pumps rotate on top of a fixed brick or stone turret. In this case, the sails drive two rag and chain pumps used to lift water from a well to the top of the turret from where it may be used to irrigate a garden or to feed garden fountains. A roller at the bottom of the suction pipes help to guide the leather plungers into the pipes. Exactly when the turret mill first appeared is not known and it may be a much older design than is generally supposed. Medieval drawings of windmills are relatively common and sometimes show what appears to be a turret rather than a post mill, but the evidence is not beyond reasonable doubt. Ramelli's mills show simple hand operated winches for pulling the cap and sails into or out of the wind, which must have been a tedious task.

An interesting application of wind power is Leonardo's roasting spit illustrated in Figure 3.5. It is driven by a rising current of flue gases in the chimney above a fire. These gases rotate a simple vane turbine placed in the chimney and its rotation drives the mechanism of the turnspit. Leonardo certainly experimented with convective air currents. By way of introduction to this device he remarks that when two equally heavy objects are suspended from the two sides of a balance, and the air below one of them is heated, the hot air in rising upwards will cause one weight to rise and the other to fall. Simple rotating vanes seem to have been reasonably well known in renaissance Europe, if only as a toy Van Belle (2005, 220). Hart (1961, 353) shows a painting from the museum at Le Mans of the Christ Child playing with a whirligig toy. It was painted in about 1460. Similarly, a portion of a stained-glass window (originally at Stoke Poges, England) also shows the Christ Child playing with a whirligig toy. Hart believed that when the string was pulled the propeller vanes rotated and the toy flew upwards, like a helicopter, but recent opinion is that the vanes rotated, making a noise, but did not fly.

The most famous example of a turbine is that described by Hero of Alexandria but, like much else that he invented, it appears to have been only a toy or curiosity. Leonardo knew of the enormous increase in volume (about 1,700 times) when water boils-off into steam and in his notebooks, he describes and sketches an apparatus to

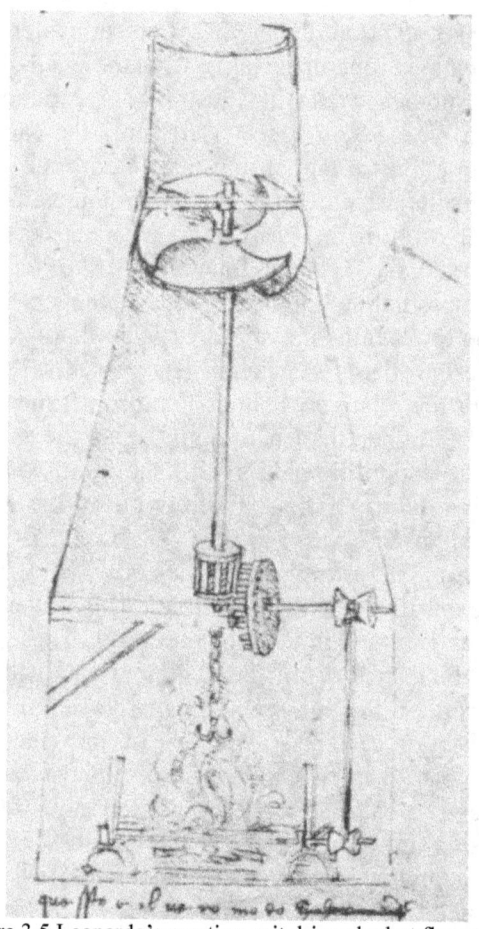
Figure 3.5 Leonardo's roasting-spit driven by hot flue gases.

measure it. He also illustrates a design for a boiler, in the shape of a man's head with a jet of steam emerging from his mouth, see Figure 3.6 (top), but he seems not to have put this to any practical use. Certainly, a boiler without an adequate safety valve would be a very dangerous piece of equipment, so if Leonardo did build such a device he might well have been discouraged.

Nevertheless, such boilers were built as the 15th century example in the Correr Museum testifies, Figure 3.6 (bottom left), and Branca in 1629 illustrates a possible application of a steam jet to drive an impulse turbine and operate a stamp mill, Figure 3.6 (bottom right).

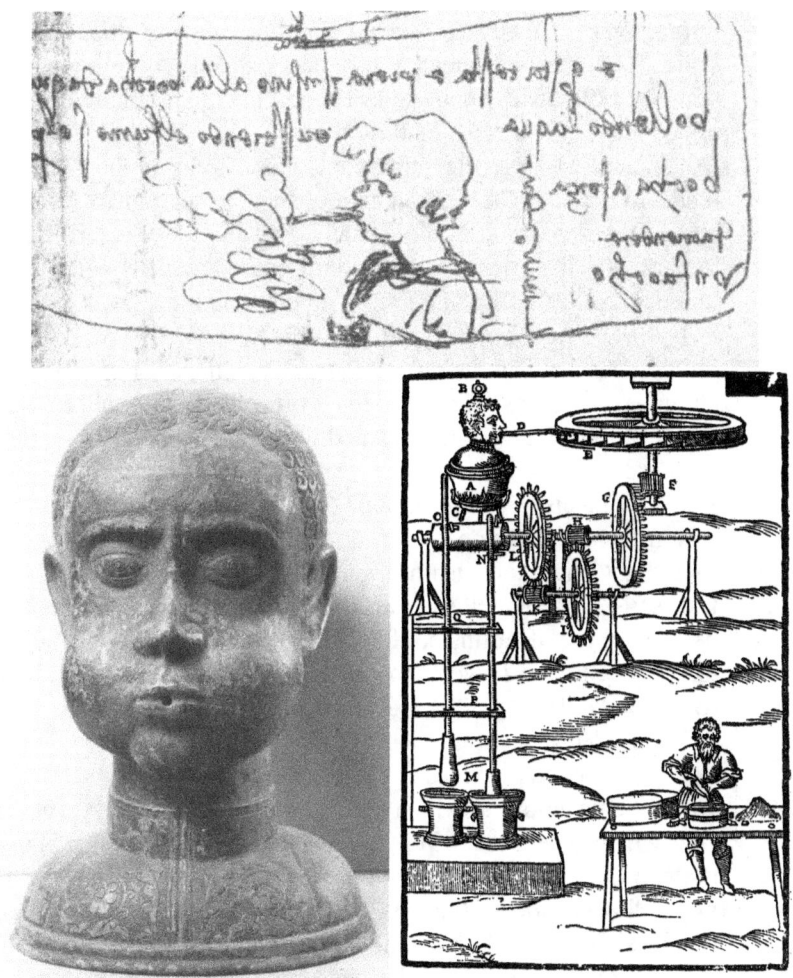

Figure 3.6 Top: Leonardo's steam jet generator.
Bottom left: a steam jet generator, 15th century.
Bottom right: an imaginary application of steam to drive a stamping-mill via a turbine.
Branca, *Le Macchine*, 1629.

A fire boils water in a pot shaped like a man's head and the steam issues from a hole in the mouth. This jet impinges onto a vertical-axis wheel that drives a simple camshaft through three crown and lantern speed reduction gears. As the cam slowly rotates it lifts heavy stamps and lets them falls into the buckets to crush their contents (ore). It is unlikely that this machine was ever built and it is usually considered impractical. However, a steam pressure of only about 2 bar in the boiler would be sufficient to produce a steam jet having a velocity of the speed of sound, about 480 m/s, and having a kinetic energy flux of about 3 kW for a hole of only 1 cm^2 area. Probably this would be insufficient to drive the stamps but there is no doubt that it could be made to work. There is an interesting connection to Leonardo here because both Branca's boiler and that in the Correr Museum are remarkably like that drawn and discussed by Leonardo. It seems inconceivable that Leonardo's notebooks had not been seen and read. Reti (1956) in a paper quoted by Hart (1961, 297) is firmly of this opinion and he demonstrates Leonardo's connections with several later engineers, such as Cardan, della Porta, de Caus, and Branca, all of whom described various steam devices in language remarkably like Leonardo's.

Francis Bacon in 1662 carried out experiments on model windmills and developed a crude theory of how the sails worked. He says that "The wind rushing against the machine is compressed by the four sails and compelled to make a passage through the four openings between them. But this confinement it does not willingly submit to; so that it begins, as it were, to jog the sides of the sails and turn them round, as children's toys are set in motion and turned by the finger", see Hills (1994, 86)

"If the sails were stretched out equally [zero weather angle] it would be uncertain to which side they would incline. As, however, the side which meets the wind [the leading edge] throws off the force of the wind to the lower side, and thence through the vacant intervals; and as the lower side is like the palm of the hand or the sail of a ship receives the wind, the rotation forthwith commences with that part. But it should be observed that the origin of the motion is not from the first impulse (that which is made in the front), but from the lateral impulse after compression has taken place."

All this is reasonable enough but his understanding was incomplete and like so many theories, it did not prove particularly useful in improving performance. Of the three experiments he carried out, two reduced the windmill's performance but the third showed some improvement. Bacon declares: "I have made several trials and experiments for increasing this motion, ...contriving imitation of the motion by means of paper sails and the wind from a pair of bellows. Accordingly, to the lower side of the sail I fastened an additional fold, turned away from the wind, in order that the wind being now directed from the side might have a larger surface to strike against. But this did no good --. At some distance behind the sails, and the whole breadth of their diameter, I placed obstacles in order that the wind, being more compressed, might strike with greater force; but this did more harm than good, as the repercussion deadened the primary motion. Again, I made the sails double their former width in order to compress the wind more and the lateral percussion stronger. This at least was completely successful, for the sails were turned by a much gentler blast and revolved much faster."

Bennett and Elton (1898, II, 260) quote a slightly fuller version of this passage. The reason why Bacon's broad sails worked was not that they compressed the air more (the pressure rise on the sails remained constant) but because the increased area generated a larger force. To increase the pressure, he needed to increase the wind velocity. Bacon, very sensibly, warned that if broader sails were used the windmill must be made stronger to withstand the increased wind force. He also recommended that instead of increasing the width of the sails the miller should consider increasing their number. Although mills with four sails were the most common, many were built with five, six, or even eight sails. Modern wind pumps, which require a high starting torque, usually have a continuous array of vanes (sails).

Parent (*Recherches de Mathématiques et de Physique, 1713*) showed that the drag force was proportional to the square of the wind velocity. He determined that the best weather angle (the angle between the plane of rotation and the plane of the sails) should be about 35 degrees. This was a perfectly good result if the sails did not move but it was far too steep for practice. Emerson (*Principles of Mechanics, 1754*) realised this deficiency but did not correct the mathematics. The same problem occurs in setting the sails on sailing ships. To maximise the force the sails had to be set at the optimum angle relative to the wind. Bouger (1757) made an analysis of this problem and showed that the optimum angle was given by *$2tan(\theta)=tan(\Phi)$*. Here θ is the angle between the sail and the keel (the weather angle of a windmill sail) and Φ is the angle between the wind direction and the sail. If the sail is still, then $\Phi=90°-\theta$ and the optimum angle is 35.3 degrees. However, when the sail moves at right angles to the wind the angle of incidence of the wind reduces and the optimum weather angle similarly reduces.

This created a problem in constructing sails because the sail speed at the tip is much greater than the speed near the centre and thus the angle of incidence of the wind on the sails varies continuously from the centre to the tip. The weather angle for maximum force must decrease from 35.3 degrees at the centre to a much smaller angle at the tip and results in the characteristic twisted appearance of mill sails. The French engineers Euler and d'Alembert continued theoretical work on the optimum weather angle of windmill sails and in England Smeaton in 1759 carried out experiments to optimise the sails, see Smeaton (1796). In his experiments, Smeaton did not rely on a bellows to provide a flow of air, as did Bacon, but moved the sails through still air on a rotating arm. He carried out 19 experiments on six or more sail configurations and experimented to optimise the weather angle at any sail radius.

The optimum weather angle determined by Smeaton, and some later results obtained by Templeton (1856) are shown in Table 3.1. These results were often used in designing and building windmill sails but it seems likely that millers had, for a long time, varied the weather angle along the length of their sails, so this theoretical and experimental work introduced no new practice. The optimum sail angle depends, of course, on the wind speed relative to the sails and so it was desirable to change the sail angle depending on wind conditions. Lancaster Burne et al (1945) give a very full account of the design and construction of windmill sails.

Table 3.1 Weather Angle of Windmill Sails

Sail Radius (sixths)	Angle of sail from plane of motion (Weather Angle) degrees	
	Smeaton (1759)	Templeton (1856)
1/6	18	24
1//3	19	21
½	18	18
2/3	16	14
5/6	12.5	9
1	7	3.37

Heaving the cap and sails round to face the wind was an arduous task and it is surprising an easier method of adjustment was not devised earlier, but fantails, that automatically rotate the cap and sails of a turret mill to maintain them always "in the Eye of ye Wind", did not appear until 1745 when Edmund Lee of Brock Mill near Wigan took out a patent, see Figure 3.7 (top). Lee's English patent gave a drawing but no description of the apparatus. However, Hills (1994, 92) has found a description in Lee's Dutch patent, which makes the operation clear. As Lee states "—when the wind shifts from the eye of the main sails, in the smallest degree, it immediately catcheth the back sails [the fantail] the turning of which, by a great power [reduction gear] it communicates to a travelling wheel which moves round the machine in which said back sails are fixed, being fastened to that outward part of the mill which turns round." The sails of the fantail were normally set at about 45 degrees but this later increased to about 55 degrees. The force generated for small deviations of the wind was equally small and so a very large reduction ratio was necessary to drive the mill cap and sails back into the wind; typically, this is about 2,000:1. Such a substantial reduction ratio required efficient gears and is, therefore, more suitable for cast iron gearing than the older wooden gears. Nevertheless, the fantail proved very effective and often the radius at which it operated had to be shortened to prevent the mill rotating too much with every change in wind direction. A more modern fantail is illustrated in Figure 3.7 (lower) and on the Wilton windmill, Figure 3.8. Typically, the fantail rotation turns a worm on the cap that meshes with gear teeth on the main structure, and thus rotates the cap relative to the main structure.

The most important adjustment when milling corn was the gap between the stones. If the stones turned too quickly in a good wind, then the stones might turn at about 150 rev/min and the milled grain would be thrown out by centrifugal force. The miller must then reduce the gap to ensure the grain is properly milled. If the wind dropped the stones might turn at only 60 rev/min or less and the grain would not be thrown out so quickly and may be burnt by remaining too long in the stones. To prevent this the miller had to widen the gap. This was known as *tentering* and the miller judged the quality of the flour by feeling it between thumb and finger as it left the stones. Judgements based on experience are still known as "the rule of thumb". To assist the miller in this never ending task the speed governor was introduced. Robert Hilton of Preston, Lancashire, made the first mill speed governor in 1785. It sensed speed changes using a centrifugal fan and

Figure 3.7. Top: Edmund Lee's patent windmill with fantail and self-regulating sails, 1745. Hills (1994, 93). Bottom: The usual design of fantail, Knight (1884, 2782).

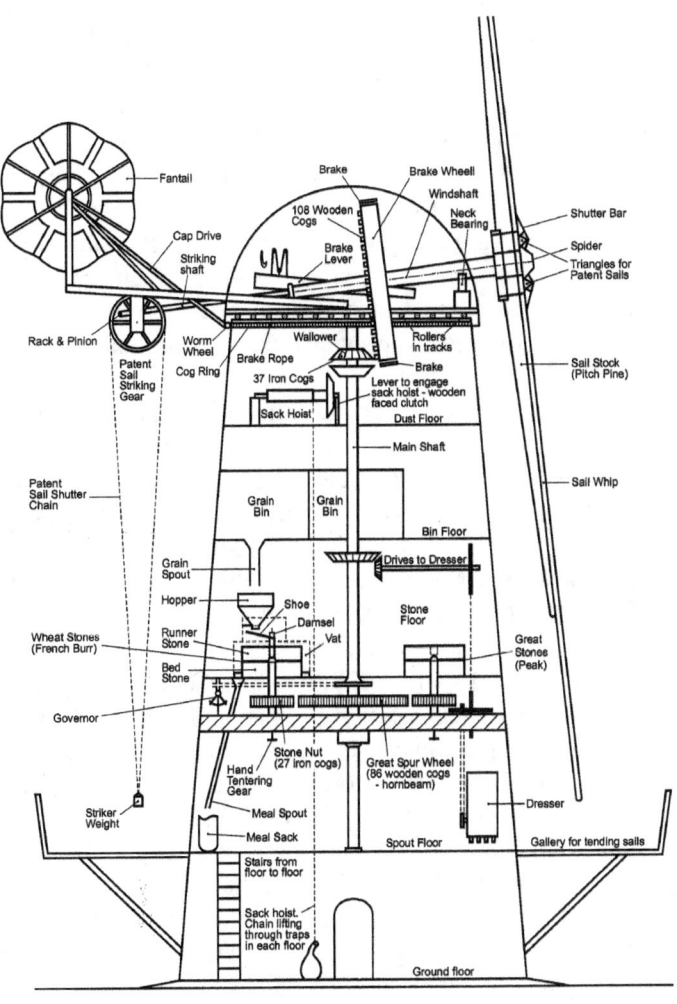

Figure 3.8 Wilton windmill. After Anon (1979).

used a tentering mechanism to adjust the size of the gap in the stones to keep the speed constant, Hills (1994, 96). Many other designs followed but the most common device was based on the centrifugal governor, as commonly used on reciprocating steam engines.

The first small windmills suitable for use on the homesteads and farms of the American west started to appear in the 1850s, Hills

(1994, 240). What was required was a simple windmill able to run unattended yet maintain its direction into the wind and run at a constant speed. During the 1850s over fifty windmill patents were filed in the USA. Usually a simple tailboard, like a weather vane, kept the sails directed at the wind. Twisted blades, popular in England for their efficiency but expensive to make, were not used, and the sail blockage ratio was increased to give a good starting torque. Many sails were mounted in a ring and a speed governing system that turned (feathered) the sail to spill the excess wind was used. Such windmills could be left unattended to pump water etc and became quite popular. Their use spread to Europe wherever a small, portable power source was required.

A method of speed regulation, appropriate for windmills driving pumps etc, was to control the torque generated by the sails either by altering their angle relative to the wind (feathering) or by altering the surface area by adjusting the area of sailcloth exposed to the wind. Numerous inventions along these lines were made, and amongst the most successful was Cubitt's invention of 1807, see Figure 3.9 (left), which came to be used on most windmills after this date. The continuous sail surface of earlier mills was replaced by a series of shutters, marked g in the figure, which could be inclined relative to the wind by movement of the rack and pinion f. This was connected through a simple linkage to another rack and pinion at d and e. The miller applied a constant torque to the pinion e by means of a weight and the force generated by the wind blowing over the shutters caused them to incline until the net force on the sails balanced the applied weight. If the wind increased then the shutters inclined further and if it decreased the shutters closed, thus the force and torque generated by the sails remained constant. To increase the torque and power the miller simply increased the weight. To stop the mill the miller removed the weight and the wind would then blow the shutters so they aligned with the wind and the sails then generated zero torque and came to rest.

Figure 3.9 (right) illustrates two further devices for controlling speed by feathering the sails. In one device, a ball S is free to slide along a rod attached to the sail. As it moves it rotates the sector R, which is connected to the head H and acts in opposition to a spring O of sufficient stiffness. As the sector R, rotates against the spring force it also rotates the sail shaft and thus alters the angle between the sail

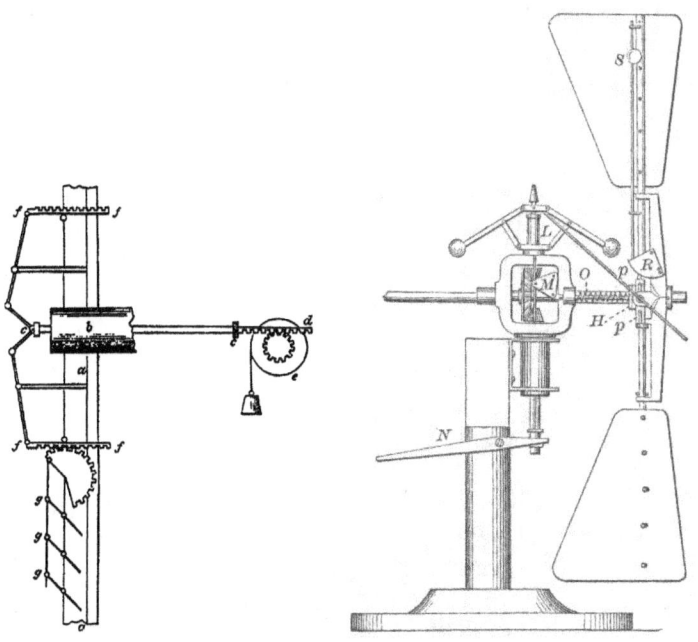

Figure 3.9 Left: Cubitt's patent speed governor, 1807. Hills (1994, 107).
Right: A later method of feathering windmill sails. Knight (1884, 2781).

and the wind. It is so arranged that as the ball S moves outwards because of an increase in speed, the sail is rotated into the wind, thus reducing the force and slowing the rotational speed.

The second device illustrated in Figure 3.9 (right) has a more conventional rotating ball governor L. This acts through the sector M and various connecting rods. As the speed increases the balls move outwards and this movement rotates the sails, decreasing the angle between the sail surface and the wind direction and so reduces the speed.

Appendices

Theoretical Power of Windmills

As wind at velocity v_0 and pressure p_0 approaches the sails of a windmill, Figure 3.10 (top), it slows down to v_0 *(1-a)*, where *a* is called the interference factor; this causes a rise in pressure. Behind the sail is a partial vacuum and this returns to p_0 at section 3 where the velocity has slowed still further to $v_0(1-b)$. As the velocity decreases the pressure rises and the area of the stream tube increases as illustrated in Figure 3.10 (top). Applying Bernoulli's theorem to section 0-1, and 2-3 gives

$$\frac{p_0}{\rho} + \frac{v_0^2}{2} = \frac{p_1}{\rho} + \frac{v_0^2}{2}(1-a)^2$$

$$\frac{p_0}{\rho} + \frac{v_0^2}{2}(1-b)^2 = \frac{p_2}{\rho} + \frac{v_0^2}{2}(1-a^2)$$

3.1

Subtracting these yields

$$p_1 - p_2 = \rho \frac{v_0^2}{2}[1-(1-b)^2]$$

3.2

Thus, the axial force acting on the sails is

$$F = (p_1 - p_2)A = \rho A \frac{v_0^2}{2}(2b-b^2)$$

3.3

The force on the sails may also be determined from the momentum equation as the product of the mass flow rate through the sails and the velocity change between 0 and 3, that is

$$F = \rho A v_0^2 (1-a)b$$

3.4

Equating 3.3 and 3.4 gives $b = 2a$, and so the power transmitted by the sails is

$$W = F v_0 (1-a) = 4a(1-a)^2 \rho A \frac{v_0^3}{2}$$

3.5

The efficiency of the sails may be defined as

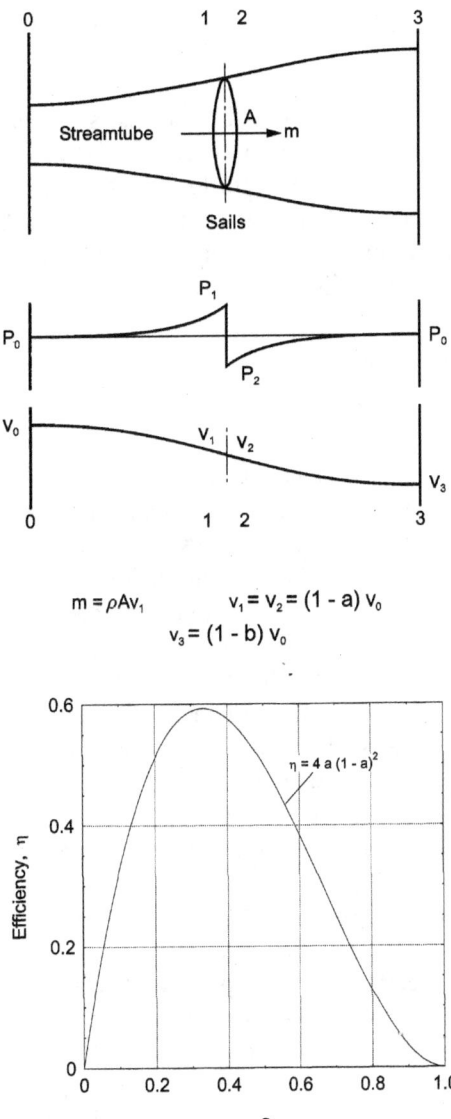

Figure 3.10 Top: air flow through a windmill (stream tube) showing pressure and velocity variation. Bottom: theoretical efficiency of windmills.

$$\eta = \frac{2W}{\rho A v^3} = 4a(1-a)^2 \qquad 3.6$$

Clearly the power developed by the sails increases with the cube of the wind velocity and is zero when $a = 0$ or when $a = 1$. At some intermediate value of the interference factor, a, the power must reach a maximum value. Differentiating Equation 3.6 and equating to zero gives maximum power when $a = 1/3$ and the maximum power generated by the sails is thus

$$W_{max} = \frac{16}{27} \rho A \frac{v_0^3}{2} \qquad 3.7$$

The maximum efficiency is $16/27 = 0.593$ and is known as the Betz limit after the German aerodynamicist who first derived it.

Figure 3.10 (bottom), shows how the efficiency varies with interference factor. Clearly, the efficiency remains quite high, above 40%, for a wide range of interference factors and the maximum efficiency occurs, as we have seen, when a = 1/3. The interference factor is small when the load on the sails is small; under these conditions the wind energy is carried off in the wake rather than used to drive the sails. On the other hand, the interference factor is large when the load on the sails is large; under these conditions, very little air flows through the sails and the pressure rise in front of the sail causes much of the wind to be deflected around the sails.

Wind Distribution

The wind speeds at which typical windmills start, operate well, or must be closed are shown in Table 3.2. Old Dutch mills required a wind speed of about 6 m/s before they would rotate. English mills of later design, on the other hand, usually started at about 3.5 m/s. The sails of Old Dutch mills needed to be reefed at about 11 m/s and furled at about 13.5 m/s. The later English mills could operate in higher winds and were partly furled at speeds of about 13 m/s and not closed until about 18 m/s. The magnitude of the wind velocity, u, is found to vary randomly about some mean value approximating to the Weibull

Table 3.2 Wind Speed and Operating Conditions of Windmills, after Hills (1994,8).

Beaufort Scale	Wind Speed m/s	Old Dutch Windmill	English Windmills
0 Calm	0-0.5		
1 Calm	0.5-1		
2 Light Air	1.5-2.6		
3 Gentle Breeze	2.6-5.1		Starts
4 Moderate Breeze	5.1-7.7	Starts	Works well
5 Fresh Breeze	7.7-10.3	Works well	Best service
6 Strong Breeze	10.3-12.9	Reef sails	Max. Service
7 Moderate Gale	12.9-18	Close down	
8 Fresh Gale	18-20.5		Sails partly furled
9 Strong Gale	20.5-23.1		Close down
10 Storm	23.1-28.3		
11 Great Storm	28.3-33.4		
12 Hurricane	>33.4		

probability distribution. Thus, the probability that the magnitude of the wind speed exceeds a wind speed, v, may be expressed as

$$p(u > v) = \exp\left(-u^k / c^k\right) \qquad 3.8$$

k is the shape factor c is the characteristic velocity and is related to the mean wind velocity by $c = v_m / [\ln(2)]^{1/k} = 1.443^{1/k} v_m$. If the distribution is Gaussian then the shape factor is $k = 2$, but in practice it is usually found that 1.7<k<2.3 and the most typical value is k=1.85. Equation 3.8 is illustrated in Figure 3.11 for shape factor k=2 and characteristic velocity c=2.5, 5, 7.5, and 10 m/s corresponding to calm, breezy, windy, and very windy locations. The figure also shows typical values of the wind speed required to start a mill, about 3.5 m/s, and the wind speed at which it becomes unsafe to continue milling on full sails, about 12.5 m/s. If the mill is in a calm area (c=2.5 m/s) then the wind speed is sufficient for only 15% of the time but is rarely too high. If the mill is in a very windy position (c=10 m/s) then it has sufficient wind for 89% of the time but for 11% of the time there is too much wind and the danger of the occasional excessive or freak wind is much higher. On these figures, the mill may run for 68% of the time in

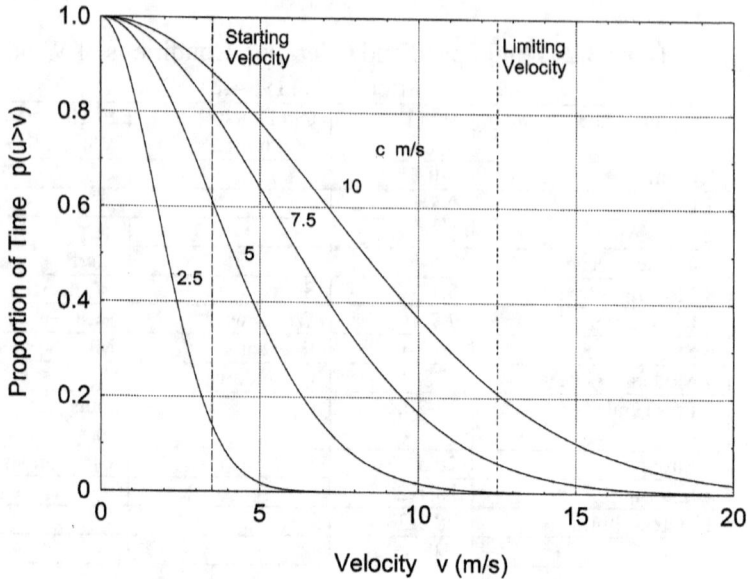

Figure 3.11 Probability of wind speed exceeding certain values.

a very windy area, 74% of its time in a moderately windy area, for 60% of its time in a breezy area and 15% of its time in a calm area. Moderately windy conditions (c=7.5 m/s) are preferred if the proportion of time that the mill is available to run is to be a maximum. If the wind speed at which the mill starts is v_1 and the speed at which it must be stopped is v_2, then the proportion of time that the mill can operate, p, is

$$p = \exp[-(v_1/c)^k] - \exp([-(v_2/c)^k] \qquad 3.9$$

It is easily shown that p is a maximum when the characteristic wind velocity is given by

$$c^k = \frac{v_2^k - v_1^k}{\ln[v_2^k / v_1^k]} \qquad 3.10$$

Modern practice is to design wind structures to withstand the extreme wind speed that occurs for a period of one hour in every 50 years. The number of hours per year is 8766 thus the wind velocity that occurs for 1/(8766x50) of the time is given by Equation 3.8 and is

$$v_{max} = 13^{1/k} c \qquad 3.11$$

If k=1.85 then v_{max}=4c=4.88v_m. Under these conditions, the pressures and forces exerted on the structure are about 24 times greater than the average and perhaps 6 times greater than when the mill is normally shut down because of high wind. It is hardly surprising, therefore, that medieval windmills were apt to be blown down every 50 years or less. It should be noted, however, that Weibull's equation, Equation 3.8, is not particularly accurate for predicting the probability of extreme gusts and modern practice is based on the equations outlined, for example, by Hassan and Sykes (1990, 11-30).

Chapter 4 Land Transport

Introduction

Transport has always been an important human activity. Early man, until about 5,000-10,000 years ago, was a nomadic hunter-gatherer; moving on with the change in the seasons, the exhaustion of local supplies, or the movement of the herds he hunted. In the beginning, the family carried their own possessions, but at some point, animals were captured and domesticated, and some were found useful as pack animals. The exact time and place at which this first occurred is not known. Certainly, by 8500 BC and 7000 BC the paddle and the canoe respectively are found in Europe. It is likely that water transport had been in existence much earlier for fishhooks and harpoons of Late Palaeolithic dates are frequently found. In the Neolithic, with the invention of farming, a more settled way of life was necessary and eventually, from about 5000 BC onwards villages, towns, and cities developed. This more sedentary life might be expected to reduce the need for transport but the converse was true for raw materials and goods now had to be brought long distances, from where they occurred naturally or were made, to the towns and villages where they were consumed. Consequently, by the fourth millennium BC, transport systems had developed and animals were in use to carry goods on their backs or to pull carts.

Prehistory and the Ancient Civilisations

Piggott (1983, 14, 96-135) notes that wheeled vehicles first appear around the 4th millennium BC within the area from the Rhine to the Tigris, and they did not appear in India until the third millennium and in Egypt and China until the second millennium BC. Before the first historical contacts they were unknown in South East Asia, Africa south of the Sahara, Australasia, Polynesia, and the Americas. The critical factor was the availability of animal power (domestic cattle or the horse). Evidence of wheeled vehicles in the Near East consists of pictographic signs on inscribed clay tablets found in Uruk, Mesopotamia, around 3200-3100 BC, Figure 4.1, Littauer and Crouwel (1979). Inherently, the wheel is a very efficient means of transport but it is made inefficient by the softness of terrain over which it must travel. The earliest wheels were solid wooden discs, several of which have been recovered from bogs.

Figure 4.1 Mesopotamian signs representing wheeled vehicles.
After Littauer and Crouwel (1979, Fig.1)

These were not cut from the circular section of tree trunks for such a construction is inherently weak. As Piggott points out, the earliest sawn wood planks in Europe are of the 6th century BC, which dates the invention of the saw, and consequently the earliest wooden wheels were made by splitting logs lengthways with wedges to form planks. These were then shaped with an adze.

Donkeys and oxen were the traditional beasts of burden. They hauled ploughs and carts, and donkey caravans carried trade to distant places. The military side of the "Royal Standard of Ur", Figure 4.2, shows several four-wheeled battle cars, the precursor of the two-wheeled chariot, each pulled by four donkeys, or onagers. There was room for a driver with warrior standing behind him.

Figure 4.2 Battle cars (early chariots) from the military side of the Royal Standard of Ur, about 2500 BC. British Museum.

Two types of two-wheeled vehicles, each carrying a single person, also developed after 3000 BC probably for military purposes. In the straddle car, the driver sat or stood on the axle and astride the draught pole. In the other, the platform car, he was seated. All vehicles had composite, tripartite disk wheels, apparently revolving on fixed axles, and draught beasts were paired either side of the pole and under a yoke, first developed for bovids, and held on by neck straps. Four beasts abreast were sometimes used with the outside beasts attached but not under yoke.

The axles of the four-wheeled battle cars were fixed, so the car turned with difficulty by dragging the wheels over the ground, causing rapid wear. Two-wheeled vehicles, if the wheels were free to rotate on fixed axles, rotated independently when turning and did not skid or wear so rapidly. Later in the 3rd millennium, disk wheels, having a metal tyre held on by clamps, first appeared, no doubt to reduce the excessive wear rates.

The four-wheeled battle car gave way, after 2000 BC, to the chariot: a light, two-wheeled, horse drawn vehicle with spoked-wheels. The wheel height never seems to have exceeded 1m. Evidently, the axles were lengthened to provide a wider wheelbase, which increased stability on slopes and during fast turns and permitted a crew of two, a driver and a warrior, with the military advantage that this implied. The earliest chariot wheels had four spokes, but in Syria, for example, six, eight, and even nine spokes occur. The axle position that minimises the draught is directly beneath the load, but for travel over soft surfaces an axle positioned towards the rear is better. Certainly, Egyptian chariots usually had a rear axle, more suitable for soft terrain, whereas Greek chariots had a central axle, suitable for hard surfaces.

The light, fast, horse drawn chariot, reached its zenith in the late 2nd millennium and is best seen in surviving examples from Egyptian tombs and paintings. The horse was not native to Egypt and Mesopotamia, but had to be imported from the steppes to the north and east. Burford (1960, 7) says that no horse remains have been found in Crete before about 1700 BC and none on the Greek mainland before about 1900 BC. It is a remarkable fact that in Egypt, wheeled vehicles are not found until the Second Intermediate period (1782-1570 BC).

Horses, once introduced, were commonly used in Egypt. The chariots found in Tutankhamen's tomb (1347-1338 BC) were lavishly decorated and are the oldest complete wheeled vehicles to have survived from antiquity. A similar chariot is displayed in Florence, Figure 4.3. Heat was used to shape and bend the poles and felloes and the floor is of interlaced leather thongs. The yoke rests on a saddle on the horse's back and is held by a girth strap and breast harness.

Assyrian palace reliefs of the early 1st millennium illustrate the further development, and ultimate demise, of chariots. Four wheeled vehicles are only rarely illustrated and seem not to have been used very

Figure 4.3 Egyptian chariot of wood, bronze and leather.
Museo Archaeologico, Florence.

much, probably because of the excessive and inconvenient wear rate of their tyres. In Cyprus, the remains of carts have been found and these had revolving axles protected by iron bearing shoes. Wheels fixed on revolving axles means that the wheels must skid sideways during turns, so this is a rather poor arrangement.

The relief Ashurbanipal, Figure 4.4, illustrates that chariots had, by the 7th century BC, become larger to accommodate a crew of four: a driver, a warrior, and two shield bearers. Clearly the fast, light chariot had become vulnerable and needed to be defended and protected, probably by attacks by increasingly important mounted troops, for riding had developed to the stage where warriors could fight from a fast-moving horse. By the late 1st millennium, chariots had lost their original role and failed to gain a new one.

An intriguing problem of prehistoric land transport is that of transporting the stone megaliths used in the construction of the numerous stone circles and tombs that abound in Western Europe. There are upwards of 40,000 such monuments. The most important and famous are at Stonehenge, Avebury, and Carnac in Brittany. The Grand Menhir Brusé, in Brittany weights about 380 tonnes and is the largest European monolith. Davidson (1961, 11-16) put forward some interesting ideas on the movement of 60 ton statues in early Assyria and Egypt, Figure 4.5, copied from a grotto at El Bersheh. It shows 172 men pulling an alabaster colossus. Davidson rejects the idea that the statue is being moved on a

Figure 4.4. 7th century BC chariot of Ashurbanipal.

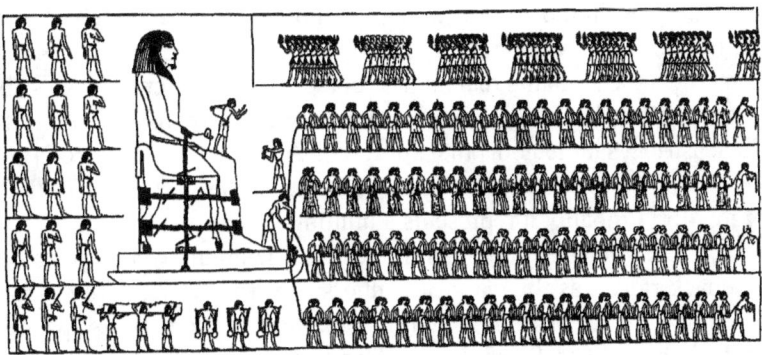

Figure 4.5. Transporting an Egyptian colossus with 172 men, about 1880 BC. After Davidson (1961). The hieroglyphics have been removed for clarity.

sledge or on wooden rollers over compacted sand. Rollers are ill adapted to move heavy weights in the direction of the pull and tend to move in a curve rather than a straight line. The sledge runners appear to have dimensions of 4.9 m x 0.46 m and the pressure exerted on the terrain by the 60 ton load is thus about 1.35 bar. Davidson notes that this is the normal pressure for modern tank tracks and is suitable for movement over sand without sinking. Men below the statue carry heavy timbers and pots of liquid for lubrication. Davidson believes that the wooden sledge runners slide on these heavy timbers and that the wooden sliding surfaces are lubricated. The coefficient of friction for hard wood on hard wood is 0.16, and assuming each man can pull 120 lb (534 N), Davidson estimated the number of men required is 179. This is not too different from the number illustrated, however, 534 N is rather high for most men and 300 N is more realistic. The coefficient of friction of lubricated oak sliding on oak is 0.072, Avallone and Burmeister (1996, 3.24), so a more realistic estimate of the number of men required is 144 (2.4 men per tonne). But even a slight slope would create problems. Medieval and early modern practice in moving heavy cannon was to allow about 20 men per tonne.

Riding

An equine skeleton found buried beneath a human in southern Mesopotamia, and dating from before 3000 BC, provides indirect evidence of riding. No doubt riding developed from the use of pack animals. Later in the 3rd millennium, equids were ridden bareback, and control was by a line and nose ring adopted from bovine practice. Riders at first sat at the animal's rear on its croup or loins, or even sideways. By the early 2nd millennium, horses were more common and were mostly ridden astride; saddlecloths first appeared. A Bronze Age image of a man mounted on a horse is engraved on a rock wall in the Val Camonica, Italy, Figure 4.6. Probably it represents a hunting scene.

The Persians, dominant in the 1st millennium, were a people of horseback riders, as the classical authors confirm. Horses were ridden astride, with the use of often elaborate, and sometimes padded, saddlecloths. Control was by bitted bridles. These riders used both the spear and the bow from horseback, and employed the famous Parthian shot, in which a rider can turn in his seat and fire a backward arrow when pursued. Horses were also ridden in the hunt and in the famous long distant messenger service inaugurated by Darius I.

The Persians, in addition to being great equestrians, were responsible for the introduction of the dromedary and the camel, both as a riding animal and as a beast of burden, Forbes[2] (1955, 134). The camel was first domesticated in central Asia and Arabia in the late Neolithic. It was first used for mountainous terrain in about 1100 BC and the dromedary was adopted for desert travel about 700 BC.

Figure 4.6. Top: rock engraving of a horse, rider, and dog. Late Bronze Age, Val Camonica, Italy.

Greece and Rome

Wheeled carriages and carts were used but passenger carriages were considered unmanly and were used mainly by women. The typical Greek country cart or carriage of the 4th century BC had clumsy and weak wheel construction, although their chariots, to judge from the many surviving drawings on Greek pottery, were light, elegant, and strong. The axle position of Greek chariots varied but was most frequently directly beneath the load and is the most suitable position for chariots driven over hard ground. Horse breeding was a considerable source of wealth. Riding for pleasure was a favourite pastime, and horses were raced at the Olympic Games. More seriously, the cavalry was an important arm of the Greek military. The second class of Greek citizens was known as the *Hippeis* (knights) because they could afford to keep a horse and provide a mounted warrior when the need arose. Men rode with a simple cloth, and without the aid of either stirrups or saddle. Horse bits were used, and could be either mild or severe, but horseshoes were unknown. Thucydides (*Histories* 7.27.5) tells us that when "the cavalry rode out daily upon excursions to Decelea and to guard the country, their horses were either lamed by being constantly worked upon the rocky ground, or wounded by the enemy". It is, perhaps, an early example of a war being lost for want of a horseshoe.

Chariots were usually drawn by a pair of horses either side of a central pole. With four horses the additional horses were not yoked but were tethered four-abreast and drew the chariot through traces. Harnessing in file, in the modern fashion, was unknown. In about 406 BC, when the commander Abradatus transferred his allegiance to Cyrus, he demonstrated his remarkable chariot, which was pulled by eight horses abreast and harnessed to four poles, (Xenophon, *Cyropaedia*, 6.1.52). When Cyrus saw this "he conceived the idea that it was possible to make one even with eight poles, so as to move with eight yoke of oxen the lowest of his movable [siege] towers." The load was 15 talents (380 kg) per yoke, whereas the usual allowance was twenty-five talents (632 kg) per yoke of oxen. Rankine (1889, 251) states that the normal pulling load for an ox is 120 lb (540 N) so the coefficient of rolling (pulling force divided by load) is 0.175.

Building stone, required for major public works such as temples, was expensive, both to quarry and to transport. Surviving Greek inscriptions shed some light on these costs, see Burford (1960, 1-18) and Glotz (1923, 26-45) and (1965, 292). When blocks of marble had to be moved by road from the port of Epidauros up to the temple, the transport cost amounted to 42% of their purchase price. To take column drums from Pentelicon to Eleusis, the road had to be repaired, wagons had to be built, and many teams of oxen had to be hired. The 5-tonne column drums, had to be carried about 40 km. Each drum required between 20 and 33 yoke of oxen and took three days. It is not known how such large numbers were yoked to the cart. As mentioned above, the normal load for a cart pulled by a yoke of oxen was about 630 kg so the heaviest drums required 8 yoke. Cyrus, mentioned above, used 8 poles to yoke his oxen so perhaps a similar system was used. Burford, however, believes they must have been yoked in file. The average cost amounted to 342 drachmae per block, which was about two years pay for a Greek artisan. As mentioned above, Lefebvre des Noëttes thought the horse, although more powerful than the ox, was unable to exert his full force because of the choking effect of the harness then in use. This restriction did not apply to oxen.

The transport of fragile goods, says Glotz (1965, 291-295), was almost as expensive as for heavy goods. From Laciadae to Eleusis, about 20 km, a hundred tiles cost 40% of their purchase price. The cost of hauling up a hillside was immensely expensive, as the Delphic administration discovered. Four blocks of tufa, which cost 244 drachmae for cutting and 896 drachmae for carriage to the wharf, then cost 1,680 drachmae "from the sea to the temple", a rise of about 600 m. Stones costing 61 drachmae each in Corinth cost 705 drachmae on delivery at Delphi

Roman roads were markedly better than those of ancient Greece, but they did not always reduce transportation costs. In Diocletian's time, about 300 AD, the official price of a cartload of hay was 600 denari and its transport cost 20 denari per mile, which made the cost of transporting it 30 miles prohibitive, Lopez (1956, 18). Pack animals were commonly used over difficult terrain.

The need to protect the horse's hoofs when travelling over rocky ground led to the use of horseshoes. Some pre-Roman examples are known but Roman examples are more common. Shoes for both horses and mules have been found in London, one with a nail still in position, see Figure 4.7 (top). This also shows a 'hippo-sandal', which was an iron plate with raised sides and hooks at the front and rear. The underside was sometimes ridged and they have been discovered fastened to horses' hooves. Their exact purpose is not known. Iron or bronze bits are commonly discovered and carvings show that saddles were sometimes used. Stirrups remained unknown.

Carts and other vehicles were often depicted on Roman memorials, and Roman literature abounds with references. Three wheels have been unearthed in the Roman rubbish pits at Newstead, Liversidge (1973, 395-6). One has ash felloes bent into a circle and fastened with iron plates. An iron tyre was shrunk onto the outside of the wheel, that is, the tyre was heated, positioned over the assembled felloes, and cooled with water to contract and thus compress the wooden wheel. No nails were required to hold the tyre in place. Eleven willow spokes and an elm hub were turned on a lathe and the hub was strengthened and protected from wear by iron rings. An iron linch pin held the wheel to its axle. One

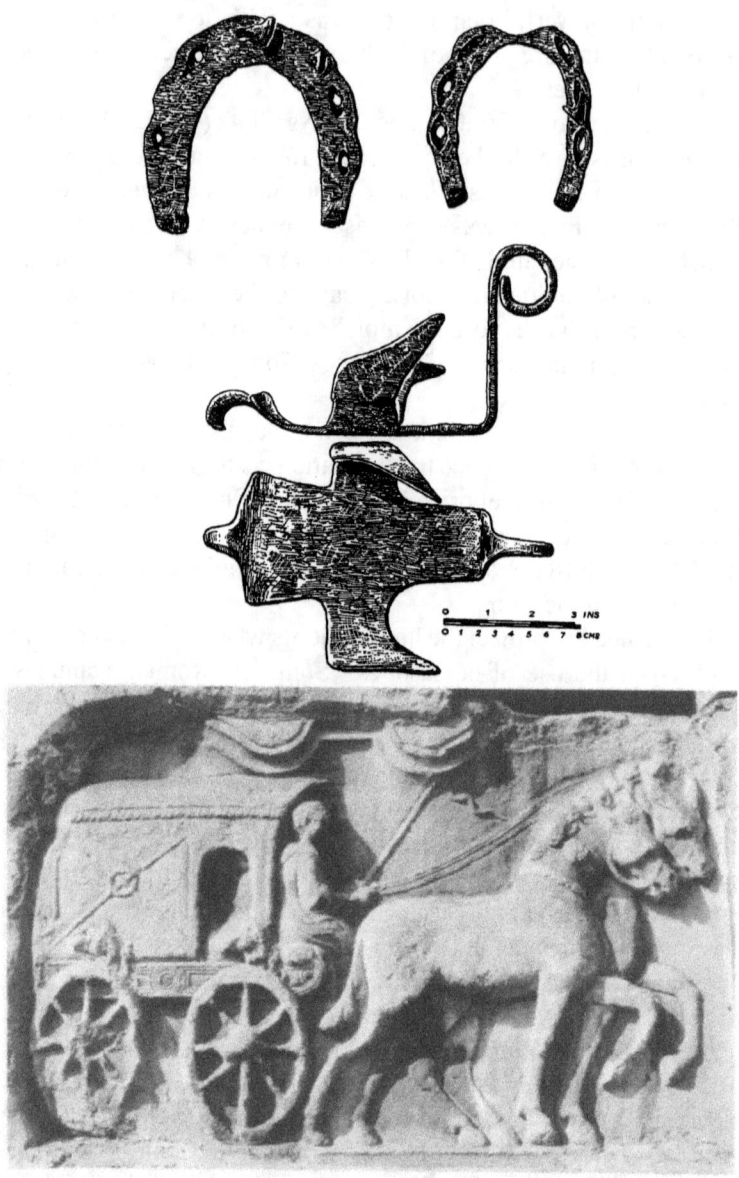

Figure 4.7 Top: Roman horse and mule shoes and the so-called hippo-sandal. Bottom: Roman covered travelling wagon (carruca), 1st or 2nd century AD, Maria Saal near Klagenfurt, Austria.

of the Newstead wheels was of heavy construction, having 12 square spokes, the ends of which projected beyond the 6 felloes and were secured with oak pegs.

British chariots, apparently, were admired. Cicero, writing to a friend in Britain, remarks "that there appears to be very little worth bringing away from Britain except the chariots" of which he wished his friend to bring him one as a pattern, Thrupp (1877, 14). The wheels of Roman chariots were usually higher than earlier designs. Those found at Pompeii were slightly dished, had ten spokes, and were 1.3 m high.

A stronger vehicle was the *rheda* or *Carruca*, Figure 4.7 (lower) which was used for carrying both goods and passengers. They were four wheeled vehicles, sometimes drawn by four horses, and carrying a driver and two or three passengers. Awnings and coachwork sometimes protected the passengers and straps were often used to suspend the seats. Heavy four-wheeled carts were much used by the Roman army and are frequently depicted on memorials. Oxen usually drew the heavier wagons. The Theodosian Code stipulates that the maximum load that could be carried in a baggage cart was 1,500 librae (about 500 kg), and is like the normal load of 632 kg per yoke specified by Xenophon. Probably this limit was to protect the road surface from damage rather than to protect the beasts. Cato states that he moved an oil mill weighing 3,560 librae (about 1,200 kg) 40 km in two days using three yoke of oxen (400 kg per yoke).

In Italy, as in Greece, animals were harnessed abreast but Lopez (1956, 18) says that in upper Italy they were sometimes yoked in file for farm labour. However, it was useless to increase the number of haulage animals so long as the vehicle itself remained small, and it was impossible to enlarge the cart so long as the roads remained narrow. Roads needed to be constructed for trade and not just for the rapid transit of messages and soldiers. From the 3rd and 4th centuries, with the barbarian invasions and the decline in the population, came the collapse of Roman power. On land, Rome gradually lost supremacy over the western provinces. It is quite remarkable that their roads survived, in some places at least, into the medieval and modern worlds.

Medieval, Renaissance, and Early Modern

Horses and packhorses, remained common, and both two-wheeled, and four wheeled carts are depicted in medieval manuscripts although little archaeological evidence survives in England from the Anglo-Saxon period. Stirrups came into use to provide the medieval knight with a firmer seat. A cart and several sleds were found among the grave goods of a Viking ship-burial at Oseberg, Norway. The four-wheeled cart, Figure 4.8, has a very ornate body and was evidently used for ceremonial purposes. The body was mounted on an ornate central rib. The wheels had spokes fastened into felloes that were thin and very deep. A few scraps of tapestry found in the same excavation illustrate both open and covered four-wheeled vehicles and suggests that, despite an almost complete lack of roads, they were in common use in Norway, and no doubt elsewhere. They are illustrated on the Bayeux Tapestry, drawn both by men and by horses.

Figure 4.8. An ornate ceremonial Viking cart found in the ship burial at Oseberg, Norway.

Land transport remained costly. Holmes (1974, 22), quotes the prior of Christ Church, Canterbury who wrote in 1323: "Know, Sir, that half of our lands are so far from us, out of this county towards Oxford,

Devonshire and elsewhere, that we must sell the grain in those parts and buy grain in this country." Money was more readily transported than grain. Nevertheless, great progress in haulage had occurred between the 10th and 14th centuries. Horseshoes, which had not been seen since the days of Rome, came into use and removed one of the objections to paved roads. In 1071 Hereward the Wake is said to have had his horse re-shod with the shoes reversed to deceive his enemies about the direction he was riding, and the outlaws Fouke and Eustace repeated the trick in other outlaw stories, Ohlgren (1998, xxvi). Equally important was the adoption of rigid horse-collars and improved harnesses, which enabled horses to pull much larger and heavier carts than hitherto. Of course, oxen could pull heavy carts, but horses were faster.

Early Modern Period

The transport of heavy weights remained as much a problem to Renaissance engineers as to those in previous ages. Ramelli (1588) suggested ten designs of cranes, seven machines for dragging heavy objects, and two machines for raising excavated earth. Some useful data on the transport of heavy ordnance by men, oxen, or by horses, was given by Thomas Smith in his "The Complete Souldier" (1628) and reproduced in Hogg (1963, 27). He suggests that one yoke of oxen is equivalent to one horse and that one horse is equivalent to six men. He allows only 270 kg per horse or per yoke of oxen and only 45 kg per man. These figures are about a half those used in Greece and Rome and probably reflect the problems in moving heavy weapons over badly damaged battlefields.

The word for a coach is similar in all European languages and is derived from the town of Kotze in Hungary where they were first built in the 15th century. There is a reference to Elizabeth 1 riding in a coach in 1559, Jackman (1962, 112), and Thrupp (1877, 38) says that the first coach in England was made in 1555 by Walter Rippon for the Earl of Rutland. Rippon also made coaches for Queen Mary in 1564 and for Queen Elizabeth. Royal patronage ensured that coach travel became fashionable towards the end of the 16th century, and even men began to give up their horses to join the ladies in their coaches. The demand for coaches by 1584 had become so great that a considerable trade in coach making had sprung-up. Typical coaches of this initial period are shown in

Figure 4.9. They were solidly built and without any springs or suspension or window glass. The wheels appear to be vertical, not dished, and thin, although they were of large diameter and the front wheels were somewhat smaller than the rear wheels to allow the coach to turn more easily. Large wheels did not sink so far into the soft roads, also illustrated in Figure 4.9, and this reduced the load on the horses. The front wheels, although smaller than the back wheels, do not appear to have sufficient room to turn without fouling the sides of the carriage. Figure 4.9 also illustrates a horse litter, still in use from antiquity. Two horses were used to pull the coaches and the load was shared equally between them by means of a whipple tree; a well-known and frequently used device that was first depicted in the Bayeux Tapestry.

Figure 4.9. Examples of 16th century coaches, and a litter.

Hackney carriages, about twenty, first appeared on London's streets in 1625-26, much to the dismay of the Thames watermen who previously had the monopoly of transport for hire in London. The profits proved to be so great that London's streets became congested and strict limits were considered. Fifty hackney coaches licensed in 1637 increased to 200 in 1652, 300 in 1654, 400 in 1662, and 700 in 1694. Private coaches, however, outnumbered the hackney carriages, and sedan chairs had become very common. By 1754, there were 4,255 four-wheeled vehicles and 2,909 two-wheeled vehicles (including chairs) and 800 public hackney carriages in London, Jackman (1962, 118, 130).

The use of coaches for transport from town to town developed. In 1649, Chamberlayne, (*The Present State of Great Britain*), says: "Besides the excellent arrangement for conveying men and letters on horseback, there is of late such an admirable commodiousness both for men and women, to travel from London to the principal towns in the country, that the like has not been seen in the world; and that is by stage coaches, wherein anyone may be transported to any place sheltered from foul weather and foul ways, free from endamaging of one's health and one's body by hard jogging or over violent motion on horseback; and this not only at the low price of a shilling for every five miles, but with such velocity and speed in one hour as the foreign post can make in one day." By 1658, stagecoaches were running to many places from London; to Devon and Cornwall, Newcastle and Edinburgh, and to Chester, Preston and Wigan, to mention but a few. None of the coaches had glass windows, and instead of springs they had leather straps, from which the body was suspended, see Figure 4.10. These did little to reduce the jolting of the occupants over the cobbled roads of the towns or to cope with the deep ruts on the mud tracks that still passed for roads all over the country.

Figure 4.10. Paris coach of 1645 with post and strap suspension, Thrupp (1877, p116).

On 1st May 1665, Pepys and some friends examined some experimental coaches that Colonel Blunt had designed on behalf of the Royal Society: "--but one did prove mighty easy (not here for me to describe, but the whole body of the coach lies upon one long spring), and we all, one after another, rid in it; and it is very fine and likely to take."

The Royal Society had supported Colonel Blunt's efforts to find a better form of suspension for coaches other than the leather strap. In September Pepys came across Colonel Blunt again. "He was in his new chariot made with springs; ---. So, for curiosity I went to try it, and up the hill to the heath [they were at Shooter's Hill, Blackheath] and over the cart ruts and found it pretty well, but not so easy as he pretends." In January 1666, the committee again went to Colonel Blunt's house. Pepys wrote: "The coachman sits astride upon a pole over the horse, which is a pretty odd thing; but it seems it is most easy for the horse, and, as they say, for the man also." Despite Colonel Blunt's efforts, the members of the Royal Society were unimpressed and the following year work on carriage suspension came to a halt.

Similar experiments were carried out in France where, by 1670, steel springs were attached to the body of a Paris Brouette. It was a further 90 years before sprung suspension was adopted for stagecoaches in Britain although they were used on private coaches. Glass windows, however, were introduced into coaches during Pepys' lifetime.

If nothing could be done to improve coach suspension, then something could be done to improve the roads. The first Act establishing a turnpike road in England was in 1663 but it was not until 1750-1770 that turnpikes became very common and thousands of Acts of Parliament established trusts. The trusts were given the right to collect a toll in exchange for providing and maintaining a road. They were local in character. Usually a group of interested parties formed a trust for a section of road, usually about 30 km in length. They raised capital for its improvement and maintenance and could charge a toll. A General Turnpike Act was passed in 1773 to speed up the parliamentary process.

By 1621, when stage wagons were coming into normal use, their excessive weight cut up the dirt roads. James I banned the use of four-wheeled wagons or carriages of more than a ton weight (1016 kg) because "excessive burdens so galled the highways, and the very foundations of bridges, that they were public nuisances". This weight is greater than that allowed in antiquity, which was about 632 kg per yoke according to Xenophon and 500 kg per yoke according to the Theodosian code.

An Act of 1662 specified that vehicles carrying for hire on the public roads should have wheels not less than 4 inches wide [10 cm] and should not be drawn by more than seven horses, or their equivalent in horses and oxen, Jackman (1962, 60). Moreover, they should not carry more than one ton between October and May nor more than 1.5 tons in the remainder of the year. The stipulation of four-inch-wide wheels was not enforced and was repealed in an Act of 1670. Jackman notes that for more than 150 years after this there were numerous complaints that narrow wheels were ruining the roads.

In 1718, a new statute recited that previous Acts had proved "wholly ineffectual" and no wagon shall be drawn by more than six horses nor cart by more than three horses. One of the main causes of bad roads was the narrow tyre fastened to the wheels with rose-head nails. If such tyres were less than 2½ inches [6.4 cm] wide then the wagon was to be drawn by no more than three horses. In 1753 wheels greater than 9 inches were encouraged by forbidding narrower wheels on highways with a penalty of the forfeiture of one horse. It was found that broad wheels could carry even larger loads and to prevent this it was enacted that the distance between the wheels, the gauge, should not exceed 5 feet six inches [1.68 m]. An example of a lumbering, heavy stage-wagon with very wide wheels and drawn by eight horses is illustrated in Figure 4.11.

Figure 4.11. An 18th century stage wagon drawn by eight horses and exhibiting very wide wheels.

Improvements in speed had to be made, and slowly, between about 1750 and 1835, coach speeds did increase. The coach speeds shown in Figure 4.12 are based on the coach travel times advertised in newspapers of the period and collected by Jackman (1962, Appendix 5). Until about 1800 coach speeds increased rather slowly from an average of about 8 km/h in 1750 to about 11 km/h in 1800, but between 1810 and 1835 coach speeds increased quite abruptly and averaged about 18 km/h in 1835. The fastest coach services attained only 15 km/h in 1750-1800. Undoubtedly some good roads existed, but it seems that the coaches were unable to make the best use of them. They were designed ruggedly for travel on more difficult roads. The slowest coaches attained only about 5 km/h, which is barely faster than walking pace and a third the speed of the fastest service. Probably they had to operate along inferior roads.

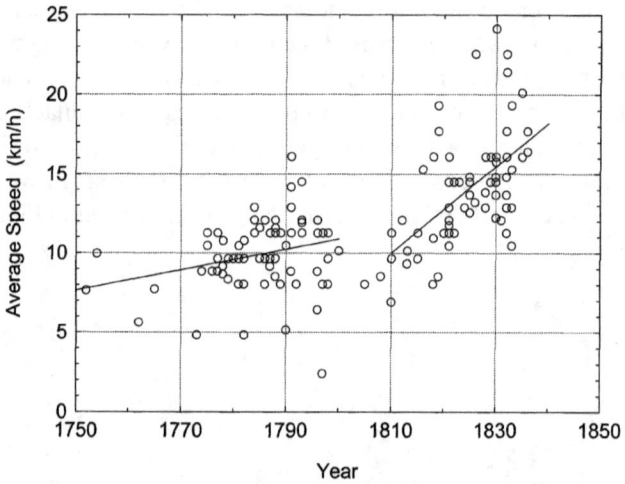

Figure 4.12. Coach speeds, 1750-1835.

Speeds in the period 1810-35 increased dramatically. The fastest coaches now attained speeds up to 24 km/h, which is attributed not only to road improvements but also to better coach design. This was the heyday of coaching, immediately before the development of the railways, but some services still contended with poor roads and attained speeds only in the range7 km/h to 10 km/h.

Improvements in the design and construction of coaches started to emerge before the end of the 18th century and reached its climax during the following century. As early as 1754 coach advertisements proclaimed the latest technical improvements. The *Ipswich Journal*, August 1754, gave notice "That a handsome Machine, with steel springs for the ease of the passengers -- began on the 8th of July 1754, to set off from Chelmsford every morning --". In the same year, the *Edinburgh Courant* stated that "The Edinburgh Stage Coach, for the better accommodation of passengers, will be altered to a new genteel two-end glass coach machine, being upon steel springs, exceedingly light and easy, to go in ten days to London in summer, and twelve in winter, every other Tuesday". The steel springs mentioned in these advertisements were either the whip spring, being an upright spring slightly curved at the top towards the body, or C-springs, and they were a considerable improvement on the rigid posts from which the leather straps were normally hung.

On 6th September 1807, *The Weekly Dispatch*, under the headline "London-Manchester in 20 Hours" reported that there were only two coaches per week between London and Manchester in 1770, but there are now twenty-seven, and what once took four and a half days now takes only thirty hours and sometimes only twenty hours by the *flying coach*."

By the end of the 18th century the construction and performance of stagecoaches had been the subject of much public controversy. They were heavy and slow; the centre of gravity was too high and the suspension was crude. Leather straps suspended the heavy body from upright C-springs attached to a beam, called a perch, which connected the front and rear axles. This arrangement is shown in Figure 4.13 (top) and elliptical springs, invented in 1805, are shown in Figure 4.13 (bottom). Philipson (1897, 36), thought that C-springs, from which the carriage was suspended by leather straps, are practically the oldest method of carriage suspension and the most perfect that has ever been introduced. Certainly, C springs and leather braces remained a popular suspension for baby carriages (perambulators), and it may be that they were comfortable for low speeds. Such coaches continued in production throughout the 19th century but they were falling into disuse. When travelling at speed, the inside passengers were subjected to a severe jolting, rather than a gentle swing, and it was considered too dangerous to carry more than one outside passenger.

Change came in 1805 with Obadiah Elliot's invention of the elliptical spring, Figure 4.13 (bottom). This enabled the box, boot, and

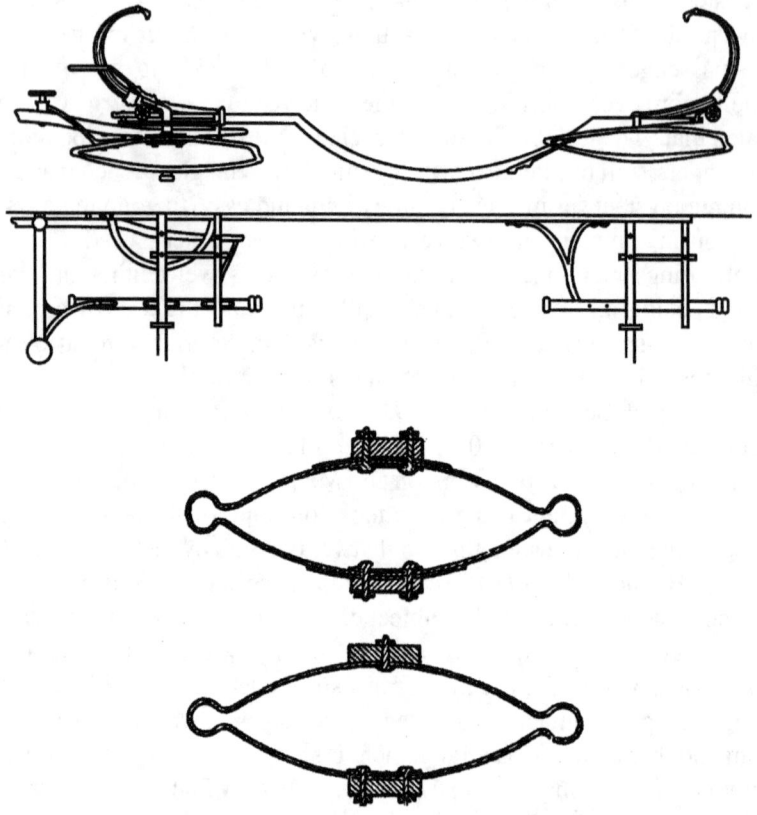

Figure 4.13. Top: Side elevation and plan of a C-spring and perch undercarriage. After Philipson (1897, 99). Bottom: Elliptical springs.

body to be to be attached directly to low-frequency horizontal springs fastened to the front and rear axles. The perch was eliminated and the weight and cost of construction was materially reduced. The comfort of the passengers increased by an order of magnitude. Thomas Hasker quickly saw the advantages of the invention and recommended to the Postmaster-General that "mail coaches should be built on the new invented plan of the body, the coachman's boot and the boot behind the coach all to be joined together and play upon and lie on horizontal springs. -- and the weight in no case to exceed 17 hundredweight [864 kg]."

The new coaches, Figure 4.14, were not only lighter but also they had a lower centre of gravity and their springs absorbed most of the ground vibration thus protecting the passengers. Consequently, they could be driven more safely and at higher speeds than their predecessors. These developments, coupled with the improved conditions of the turnpike roads, enabled the very considerable increases in speed shown in Fig.4.12 to be made. A new law of 1806 recognised these improvements by increasing the limit on outside passengers from four to ten in winter and twelve in summer. The Royal Mail coaches, however, retained a limit of three outside passengers. Thus, in addition to their increased speed, the new design of coach increased the number of passengers and, presumably, reduced the cost of transport. Elliot, with his invention of the elliptical spring, had succeeded where Colonel Blunt, 140 years earlier, had failed.

Figure 4.14 The "Liverpool Umpire", one of the new, faster, coaches. Aquatint by G. Hunt and J. Pollard after J. Moore; 1830-37. The Parker Gallery, London.

Appendices

Two-wheeled Vehicles

A two-wheeled-vehicle may be a sledge having a point contact, a wheelbarrow, a cart, a chariot or a wide range of other similar vehicles. They may be pushed or pulled manually or by a beast. A generalised two-wheeled machine is illustrated in Figure 4.15. It consists of a single-axle

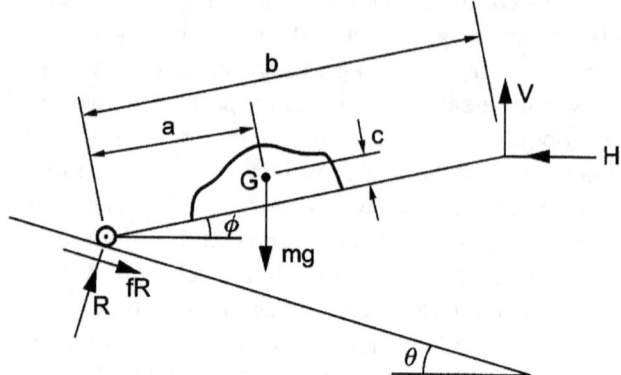

Figure 4.15. Forces and geometry of a two-wheeled vehicle ascending a hill ($\theta > 0$) at constant speed.

having one or two wheels rolling over the ground, and carries a load $W = mg$ (which includes the weight of the cart). The centre of gravity of the loaded cart is at G, distance a from the axle measured along a line from the axle to the point of loading which is at distance b from the axle. This line is at angle φ to the horizontal. The vehicle ascends a slight incline having a slope θ. If the slope is sufficiently small then $\cos\theta = 1$ and $\sin\theta = \theta$. The applied force has vertical and horizontal components V and H. The normal reaction between the wheel and the ground is R and the coefficient of rolling is μ_R. Taking moments about the axle gives the magnitude of the vertical force, V.

$$V = mg\left(\frac{a}{b} - \frac{c}{b}\tan\phi\right) - H\tan\phi \qquad 4.1$$

Equating the forces in the vertical and horizontal directions, we obtain

$$R = \frac{mg - V}{\cos\theta - \mu_R \sin\theta} \qquad H = R(\sin\theta + \mu_R \cos\theta) \qquad 4.2$$

The unknown quantities, H, V and R, may be determined from these three equations and expressed in the form

96

$$\frac{R}{mg} = \frac{1 - a/b + (c/b)\tan\phi}{\cos\theta(1-\tan\phi) - \sin\theta(\mu_R + \tan\phi)}$$

$$\frac{H}{mg} = \frac{[1 - a/b + (c/b)\tan\phi](\sin\theta + \mu_R\cos\theta)}{\cos\theta(1-\tan\phi) - \sin\theta(\mu_R + \tan\phi)} \qquad 4.3$$

$$\frac{V}{mg} = \frac{a}{b} - \frac{[1 - a/b + (c/b)\tan\phi](\sin\theta + \mu_R\cos\theta)\tan\phi}{\cos\theta(1-\tan\phi) - \sin\theta(\mu_R + \tan\phi)}$$

The value *(c/b)tanφ* is the product of two small numbers and can be ignored and for small values of the gradient θ and vehicle incline φ these equilibrium equations may be written in a simpler form, namely:

$$\frac{R}{mg} = \frac{1 - a/b}{(1-\phi) - \theta(\mu_R + \phi)}$$

$$\frac{H}{mg} = \frac{(1 - a/b)(\theta + \mu_R)}{(1-\phi) - \theta(\mu_R + \phi)} \qquad 4.4$$

$$\frac{V}{mg} = \frac{a}{b} - \frac{(1 - a/b)(\theta + \mu_R)\phi}{(1-\phi) - \theta(\mu_R + \phi)}$$

The resultant force required to propel the vehicle along the incline at constant velocity and is the vector sum of the forces H and V, namely:

$$\frac{F}{mg} = \left[\left(\frac{H}{mg}\right)^2 + \left(\frac{V}{mg}\right)^2\right]^{1/2} \qquad 4.5$$

This is the force experienced by the horse and its component in the direction of movement is required to determine power. Figure 4.16 shows how the resultant force F varies as the position of the load is moved from $a/b = 1$ to $a/b = 1$. If the coefficient of rolling is zero, then the optimum position of the load is immediately above the axle. This is the position favoured by the Chinese, by ancient Greek chariots and by most vehicles having hard wheels travelling over hard ground. For dirt roads, the coefficient of rolling is about 0.2 so the load should be positioned a little behind the axle. For travel over softer ground, where the coefficient of rolling might be quite high, the optimum position of the load is even further behind the axle and corresponds with western wheelbarrows that are designed to run over soft soil and to be stable when being loaded. For example, if the ground is very soft so the coefficient of rolling is 1.0 then the optimum position of the load is about midway between the lifting handle and the axle and the reduction in load at this position is about 30%. Drawings of the ancient Egyptian chariot confirm that the warrior and his driver normally took up a position slightly between the axle and the yoke and for travel over sand this is a much better arrangement. Drawings on Greek vases, on the other hand, show that the axle was

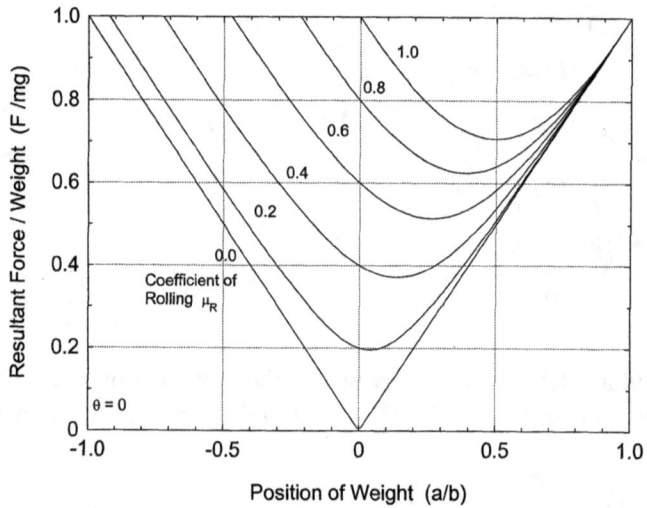

Figure 4.16. Influence of position and rolling coefficient on the resultant force required by a two-wheeled vehicle when travelling on a level surface.

often positioned directly beneath the warrior and his driver, and this is suitable for Greece's harder terrain. The wheelbarrow is most suitable for soft ground and the symmetrical design, having the load over the axle, is most suitable for hard surfaces.

Forces on Draught Beasts: Limits of Traction

A serious limit to the traction of horse drawn vehicle was set by the coefficient of friction between the horseshoe and the terrain. This was particularly critical when pulling heavy loads up a cobbled road where the coefficient of friction was low. Brigg (1893, 1034) discusses this and quotes some interesting experiments witnessed by Captain Shaw of the London Fire Brigade; the same Captain Shaw so memorably enshrined in Gilbert's lyrics for *Iolanthe*. It was found that a horse could not haul a cart up a certain hill when carrying twenty sacks each weighing one hundredweight. Two of the sacks were removed from the cart and placed on the horse's back, Figure 4.17 (top), after which the horse succeeded in climbing the hill. The following day he repeated the experiment, but this time he added two sacks to the horse without reducing the load in the cart, and again the horse ascended the hill. Brigg says this gave the horse a better "angle of draught" and was able, therefore, to get a better grip on the road surface.

Figure 4.17 (bottom) shows the forces acting on a draught beast. It is assumed that the vertical load from the weight of the shafts is small enough to be ignored and that the traces exert a force H at angle φ. The mass of the beast is m, g is local gravity, and the distance between the hooves is a, the beast's centre of gravity is b from the front hoof, and c above the ground. The draught load H is applied at height d above the ground and at distance h from the centre of gravity as indicated in the figure. The reactions normal to the gradient are R_1 and R_2 and the frictional forces parallel to the ground are S_1 and S_2.

For equilibrium of the forces acting normal to the surface and parallel to the surface, we have

$$R_1 + R_2 = mg \cos\theta + H \sin\phi$$
$$S_1 + S_2 = mg \sin\theta + H \cos\phi$$

4.6

Assume the beast slips when

$$S_1 + S_2 > f(R_1 + R_2)$$
$$mg \sin\theta + H \cos\phi > f(mg \cos\theta + H \sin\phi)$$

4.7

Re-arranging gives the maximum load that can be hauled without

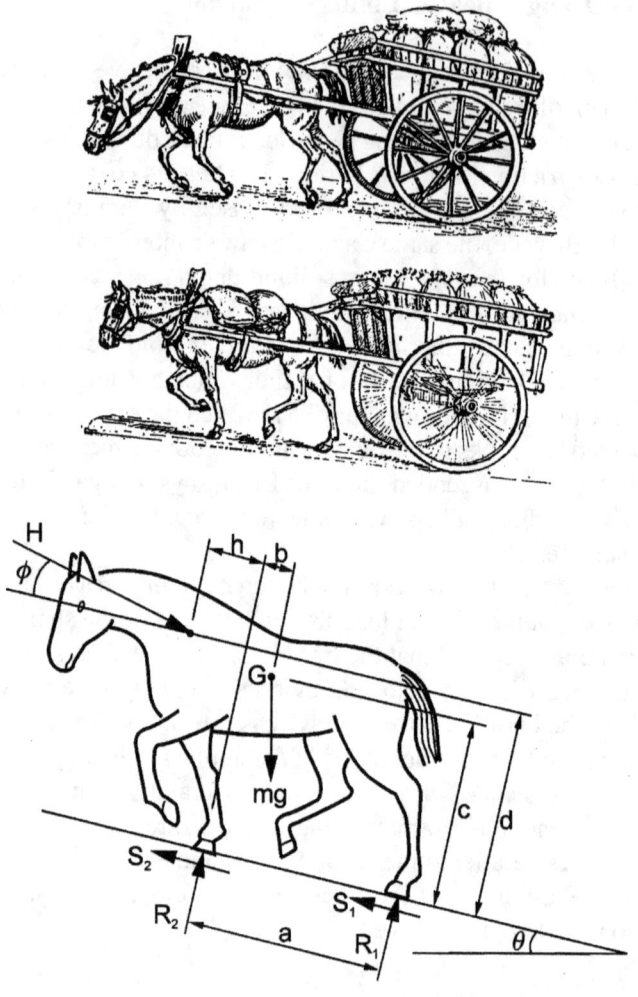

Figure 4.17. Top: removal of load (two sacks) from the cart to the horse prevents the horse slipping. Brigg (1893, 1034). Bottom: Loads acting on a draught animal.

slipping, namely

$$\frac{H}{mg} = \frac{f\cos\theta - \sin\theta}{\cos\phi - f\sin\phi} \qquad 4.8$$

Note that this result is independent of the dimensions *a, b, c, d,* and *h* and depends only on the coefficient of friction, the gradient, and the angle of the traces. The dimensions are important in fixing the magnitudes of the front and rear reactions but do not influence slip. Equation 4.8 may be approximated as

$$\frac{H}{mg} = \frac{f-\theta}{1-f\phi} \qquad 4.9$$

The normal load for a draught animal is about 10% of its weight ($H/mg=0.1$) but when ascending a hill this usually increases by about 20% when ascending 1 in 5 gradients. Consequently, when climbing it is necessary for the coefficient of friction to exceed the gradient by about 0.2 and a coefficient of friction of about 0.4 is desirable. This is reasonable on soft country roads but on cobbled or paved roads, the coefficient of friction between iron horse shoes and stone is unlikely to be so high. In these circumstances, it is necessary to reduce the haulage load per horse by increasing the number of horses or by removing part of the load and making two or more trips. Alternatively, the solution suggested in Figure 4.17 (top) may be adopted, that is, increasing the effective weight of the horse, *mg*, increases the haulage load *H*. Equation 4.9 is illustrated in Figure 4.18.

If the coefficient of friction is 0.4 then the draught animal is always able to ascend hills as steep as 20%, but if the coefficient of friction is 0.2 then the steepest hill it can climb with its load is 5%. If the coefficient of friction is very small, as, for example, when the road is icy, then the draught beasts will not be able to draw any significant load at all, even on a level road. Snow shoes can be fitted.

Wheels on Soft Surfaces

This theory is due to Bekker (1956) who imagined a rigid wheel of width *b* and radius *R*, being driven over a soft terrain, Figure 4.19 (top). The wheel sinks into the terrain to a depth z_r and compacts the soil creating a normal pressure distribution around the rim. A shear stress is also created at the rim but in this analysis, it is ignored. The weight

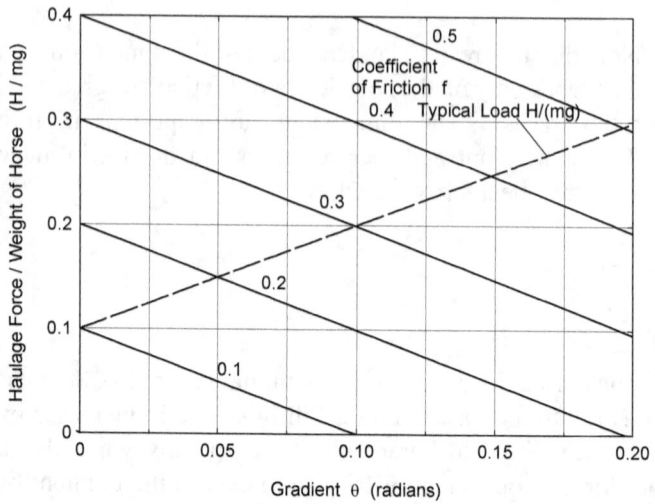

Figure 4.18. Load at which a draught beast slips when ascending a gradient.

supported by the wheel, W, is the vertical component of the pressure and the force resisting motion, F, is the horizontal component of the pressure, thus we may write

$$W = \int_0^{\theta_R} R\,b\,p \cos\theta\,d\theta \qquad F = \int_0^{\theta_r} R\,b\,p \sin\theta\,d\theta \qquad 4.10$$

If the soil is compacted by pressing a plate into it then the required pressure on the plate increases with the sinkage of the plate, z, and in general we may write

$$p = k\,z^n \qquad 4.11$$

k is the terrain stiffness coefficient (N/m^{2+n}) and n is the stiffness index. Usually $0.5 < n < 1$. Typical values of n and k for various soils are given in Figure 4.19 (bottom). Now, the relation between z and θ_r, from Figure 4.19 (top), is

$$\begin{aligned} z_R &= R(1 - \cos\theta_r) \\ z &= z_r - R(1 - \cos\theta) = R(\cos\theta - \cos\theta_r) \end{aligned} \qquad 4.12$$

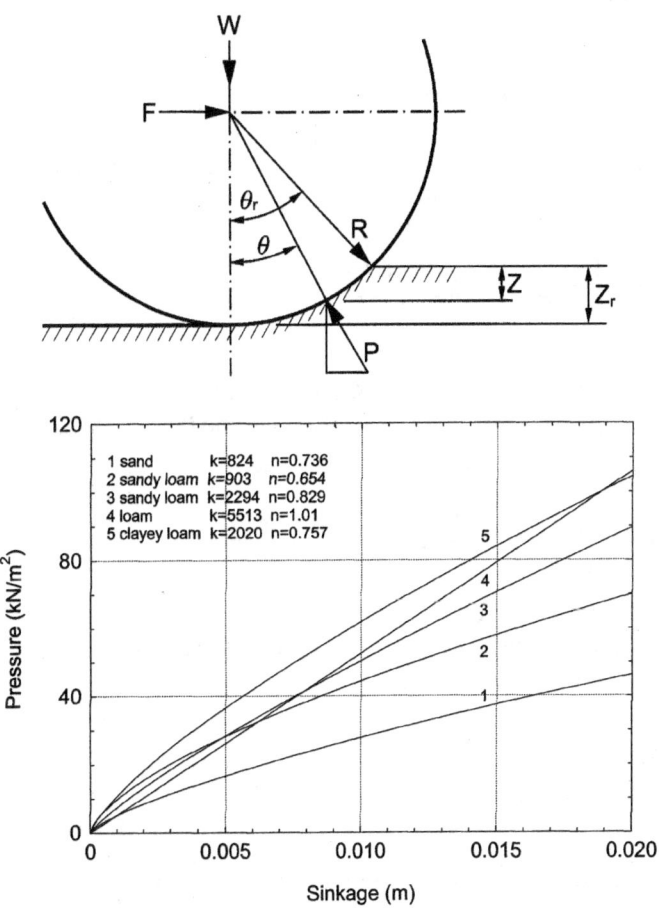

Figure 4.19 Top: simplified wheel-soil interaction model, after Wong (1993,150). Bottom: typical pressure-sinkage relations for soils for a characteristic size a=0.1m.

Substituting Equation 4.11 and 4.12 into 4.10 produces

$$W = R^{1+n} bk \int_0^{\theta_r} (\cos\theta - \cos\theta_r)^n \cos\theta \, d\theta$$

$$F = R^{1+n} bk \int_0^{\theta_r} (\cos\theta - \cos\theta_r)^n \sin\theta \, d\theta$$
4.13

Recognising that $\sin\theta \, d\theta = -d(\cos\theta)$ enables the second integral to be evaluated immediately to give

$$F = R^{1+n} bk \frac{(1-\cos\theta_r)^{n+1}}{n+1} = \frac{bk \, z_r^{n+1}}{n+1}$$
4.14

The first integral, however, is more difficult to evaluate. By numerical integration in the range $0 < \theta_r < 0.6$ radians and $0.5 < n < 1.25$, it may be shown that

$$W = 0.85 \, (2/e)^n \, bk \sqrt{D} \, z_r^{n+1/2}$$
4.15

D is the wheel diameter and $e=2.718$.

The coefficient of rolling resistance is the ratio $\mu_R = F/W$, and so from Equations 4.14 and 4.15

$$\mu_R = \frac{F}{W} = \frac{1}{0.85(n+1)} \left(\frac{e}{2}\right)^n \left(\frac{z_r}{D}\right)^{1/2}$$
4.16

z_r is defined by Equation 4.15 and may be written as

$$z_r = \left[\left(\frac{e}{2}\right)^n \frac{W}{0.85 bk \sqrt{D}} \right]^{\frac{2}{2n+1}}$$
4.17

Equation 4.16 then becomes

$$\mu_R = \frac{1}{0.85(n+1)} \left(\frac{e}{2}\right)^n \left[\left(\frac{e}{2}\right)^n \frac{W}{0.85bkD^{n+1}}\right]^{\frac{1}{2n+1}} \qquad 4.18$$

Equation 4.16, which relates the rolling resistance of the wheel to the degree to which it has sunk into the terrain is, perhaps, the easiest to understand, see Figure 4.20 (top). The value of the sinkage index, n, varies from soil to soil but is usually in the range $0.5 < n < 1$. For a sinkage of 5 cm with a 100 cm diameter wheel the coefficient of rolling resistance is about 0.2 for an average terrain. If the sinkage was only 2.5 cm or the wheel diameter was 200 cm then the coefficient of rolling reduces to about 0.13. This illustrates the advantage of hard surfaces, which avoid large sinkage and the advantage of large diameter wheels.

On ground that has been compressed by the passage of much traffic the index n is usually close to unity and the sinkage coefficient is very high, typically 40×10^6 N/m^3 or more. This greatly reduces the coefficient of rolling and is the reason why roads are compacted, using heavy road rollers, before use. The coefficient of rolling, in this case is

$$\mu_R = 0.956 \left(\frac{W}{k_u bD^2}\right)^{1/3} \qquad 4.19$$

This equation is illustrated in Figure 4.20 (bottom) for wheel loads in the range up to 100 kN/m width of wheel, and for wheel diameters of 0.25-2.0 m. The considerable advantage of wide, large diameter wheels is demonstrated.

Bekker says that his theory gives acceptable predictions of rolling resistance for moderate sinkages such that $z_r/D < 0.15$ and it appears that the theory is good for coefficients of rolling less than about 0.35. Wong (1993, 153) shows experimental evidence that the normal pressure beneath the wheel is different from that assumed by Bekker in that the maximum pressure occurs forward of bottom-dead-centre and consequently Bekker's theory underestimates the rolling resistance. In soft terrain, where the sinkage is very deep, Bekker says that the build-up of soil ahead of the wheel, ignored in his theory, creates an additional bulldozing resistance.

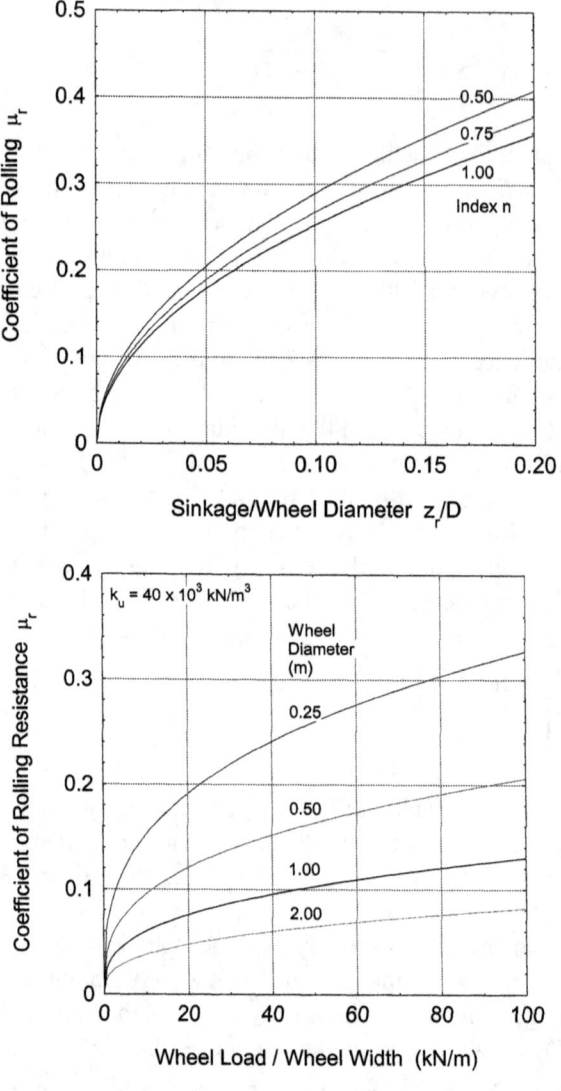

Figure 4.20 Top: relation between rolling resistance and sinkage of rigid wheels
Bottom: influence of wheel load and diameter on rolling resistance.

Wheels on Hard Surfaces

If the wheel bearing has a diameter d, the wheel diameter is D, and the coefficient of sliding friction between the wheel bearing and the axle is f, then the coefficient of rolling of the wheel, when carrying a light load on rigid smooth surfaces, is

$$\mu_R = f \frac{d}{D} \qquad 4.20$$

If we take the coefficient of friction at the bearing to be 0.2 and the ratio d/D to be 0.05 then the coefficient of rolling of a wheel, due to bearing friction, is very small, about 0.01.

When a perfectly rigid wheel contacts an uneven, rigid road surface, as in Figure 4.21, then contact occurs at the two points marked A and B. As the centre of the wheel moves it rotates about the single contact point B until contact is made at point C. Thus, the centre of the axle moves in a series of small arcs and if the vehicle is rigidly sprung then the centre of gravity of the vehicle also moves in a series of arcs. Now the change in potential energy for each arc is $mg\Delta z$ where m is the mass of the vehicle, g is local gravity and Δz is the change in height. This potential energy is dissipated in elastic vibrations and heat when the wheel falls from its maximum height to its lowest height and the wheel must be lifted by a horizontal force, F, over the next peak. The work done by this force is equal to the loss of potential energy, thus

$$FL/2 = mg\Delta z$$

$$\mu_R = \frac{F}{mg} = \frac{2\Delta z}{L} \qquad 4.21$$

Including the bearing friction, Equation 4.21 becomes

$$\mu_R = \frac{2\Delta z}{L} + f\frac{d}{D} \qquad 4.22$$

It is easily shown from the geometry that the change in height is $\Delta z = L^2/4D$ if Δz is insignificant compared with D, the diameter of the wheel, and L is the characteristic distance between bumps in the road surface. Thus, the rolling coefficient caused by the unevenness in the road surface is

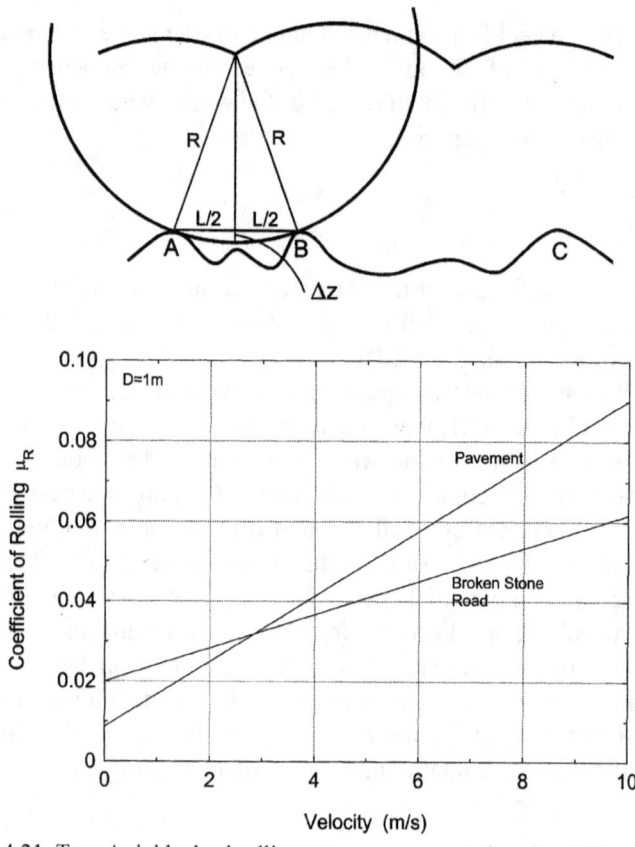

Figure 4.21. Top: A rigid wheel rolling over an uneven road surface. The centre of the axle moves in a series of short arcs. Bottom: coefficient of rolling for rigid cartwheels running over metalled roads.

$$\mu_R = \frac{L}{2D} + f\frac{d}{D} \qquad 4.23$$

Unlike rigid wheels driven across soft terrain, the rolling coefficient for wheels on rigid roads is independent of the width of the wheel and of the load applied to the wheel. Thus, if the characteristic length of the road humps is 0.1 m and the wheel diameter is 1 m then the coefficient of rolling is 0.05 regardless of wheel width or load. This value, of course, must be added to the coefficient generated by friction at the axle bearing (about 0.01). These two values have been obtained assuming the motion of the wheel is very slow and travelling over a horizontal road. If the effects of speed and gradient are also included, and bearing friction ignored, then we may use the equation:

$$\mu_R = \frac{a_0 + b_0 v}{D} + \theta \qquad 4.24$$

a_0 and b_0 are constants whose values depends on the nature of the road, D is the wheel diameter (m), v is the axle velocity (m/s), and θ is the slope of the hill (radians). Rankine (1889, 242-3) quotes an equivalent equation from which the constants, a_0 and b_0 may be determined for distinct types of road. For pavement $a_0 = 0.0086$ m and $b_0 = 0.00817$ s, whereas for a good, broken stone road $a_0 = 0.02$ m and $b_0 = .00417$s. Equation 4.24 is plotted in Figure 4.21 (bottom).

Rankine also states that typical values of the coefficients of rolling, at low speeds, are 0.14 for a sandy or gravelly ground, 0.07 for a gravel road, 0.02 to 0.03 for a broken stone road, and 0.015 for pavement. This suggests that broken stone roads are better than pavement for speeds more than about 3 m/s (10.8 km/h).

Power and Maximum Velocity

The force required to haul a load uphill is

$$H = mg(\sin\theta + \mu_R \cos\theta) \qquad 4.54$$

Where m is the mass, g is local gravity, θ is the gradient, and μ_R is the coefficient of rolling resistance. The maximum velocity, v_∞, occurs when driving force and the drag force are equal. If aerodynamic drag is small, then:

$$\frac{W}{v_\infty} = mg(\sin\theta + \mu_R \cos\theta) = mg\left(\sin\theta + \frac{a_0 + b_0 v_\infty}{D}\cos\theta\right) \qquad 4.26$$

109

Where W is the applied power. Equation 4.26 is a quadratic equation in terms of velocity and has the solution

$$v_\infty = \sqrt{\left(\frac{\theta D + a_0}{2 b_0}\right)^2 + \frac{WD}{mgb_0}} - \frac{\theta D + a_0}{2 b_0} \qquad 4.27$$

The positive root is taken. The power to load ratio required to pull a vehicle along a good gravel road ($a_0 = 0.02$ m, $b_0 = 0.00417$ s) is shown in Figure 4.22 (top), and Figure 4.22 (bottom) shows the effect of wheel diameter and gradient on velocity.

The typical power to load ratio for a heavily laden wagon is about 0.1 W/N and such a wagon might attain about 3 m/s (10.8 km/h) on a level road. To reach higher speeds the number of horses must be increased, the load must be reduced, the wheel diameter must be increased, or the quality of the road improved. The advantage of large diameter wheels is clear. For a level road surface, small wheels of about 0.25 m diameter enable a speed of 1 m/s to be attained, but very large wheels of 2 m diameter reduce the rolling resistance so a speed of about 5 m/s can be reached.

Steering

The normal arrangement for towed vehicles is that the axles are fastened rigidly, or through springs, to the main body, and the wheels are mounted so they are free to rotate independently on the axles. This arrangement allows a two-wheeled vehicle to turn very easily without slipping or skidding and the need for a steering mechanism, in this case, does not arise. However, if the front and rear axles of a four-wheeled vehicle are fixed the wheels cannot turn without being dragged because the wheels do not share a common centre of rotation, see Figure 4.23 (left). To overcome this problem a simple steering gear was introduced. Usually, the complete front axle and wheels were located on a central vertical pivot about which it could rotate. The imaginary point at which the centre lines of the two axles crossed defines the instantaneous centre about which the whole vehicle rotates. Naturally, the wheels nearest to the instantaneous centre rotate slowest and those further away rotate fastest, but they are free to do so.

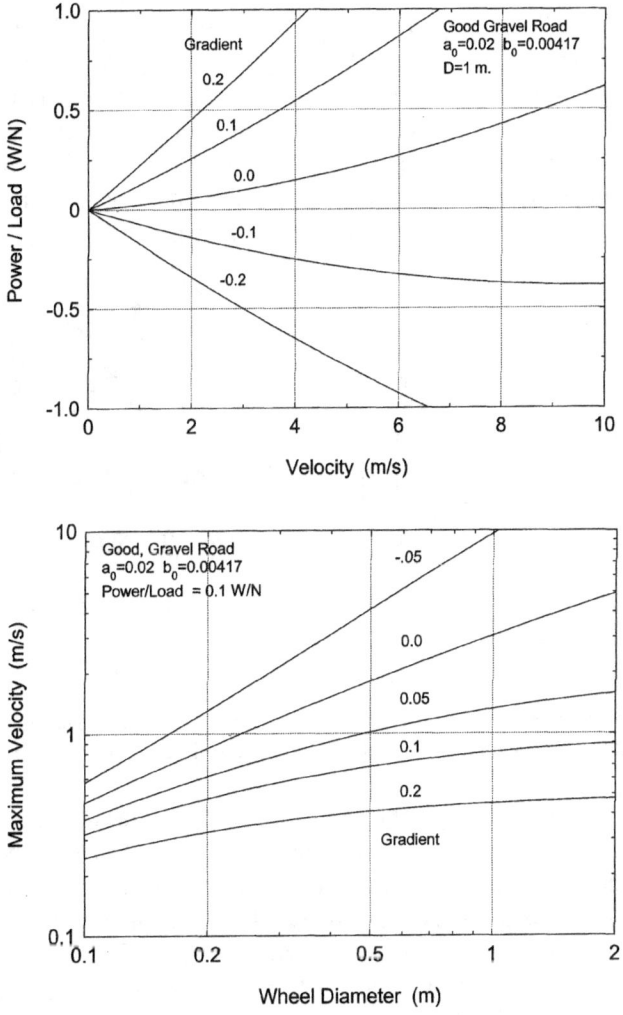

Figure 4.22. Top: power to load ratio and velocity on a good gravel road. Bottom: Maximum velocity and wheel diameter for horse drawn vehicles travelling over a good gravel road.

A typical fore-carriage for a pair of horses is illustrated in Figure 4.23 (right). For single-horse carriages they were sometimes made entirely of iron, but on two-horse carriages they vibrated and rattled and were easily bent so generally a mixture of wood and iron was preferred.

The axle-bed was usually compassed, that is, curved forward in the middle so the centre line of the main bolt was 2 to 3 inches [50-75 mm] in front of the axle. By doing this the space required by the wheels when turning was reduced and the size of the wheel-arch reduced.
Philipson (1897, 103) says that this must be kept within reasonable limits or the carriage is liable to be upset in turning because the fore-carriage is then so much further from under the body, and if there happens to be greater weight at that side then the danger is increased.

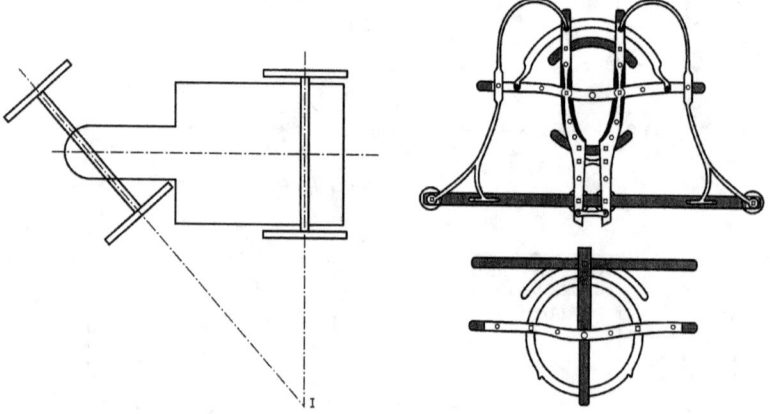

Figure 4.23. Left: steering geometry for four-wheeled vehicles.
Right: fore-carriage with close futchells and fixed splinter bar for a pair of horses.

This system worked well and its only disadvantage was that movement of the complete axle took up too much space and needed a large wheel-arch.
To reduce the wheel space required when turning on a small radius it was common practise to use steering wheels having a smaller diameter than the rear wheels. This, of course, increased the rolling resistance and caused the wheels to sink deeper into soft surfaces. Figure 4.24 (top) shows a carriage with these features. The main body is raised high above the axles on springs. The front steering wheels are smaller than the rear wheels and there is an arch in the bodywork to accommodate the wheels during tight turns. The wheels are dished and lean outwards so that the load carrying spokes are vertical as they pass under the axles, and the wheels are more able to withstand side loads, see Figure 4.23 (bottom).

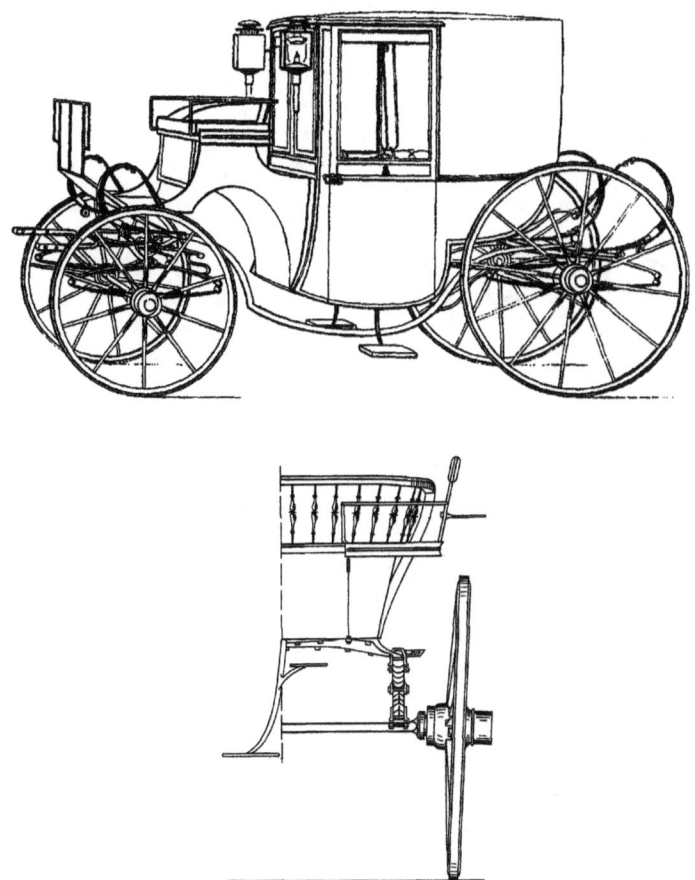

Figure 4.24. Top: a single Broughham, on C springs and under-springs with a perch, high body, wheel arch, small steering wheels, dished wheels, and sprung axles. Bottom: dished wheels to strengthen the wheels against side forces.

Chapter 5 Water Transport

Introduction

It need hardly be said that water is essential to human life and that the earliest settlements and civilisations developed close to a copious supply. The rivers Tigris and Euphrates, Nile, Jordan, Indus, Yellow River and Yang-tzi watered the earliest civilisations and without them it is doubtful that such civilisations could have developed. The importance of water is manifest. Given this affinity it is hardly surprising that water transport developed very early, and the ease with which heavy loads could be moved, when floated, gave water transport a permanent advantage over land transport. It was, and remains, the most efficient means of moving loads, particularly heavy loads.

Ancient Civilisations

In pre-dynastic Egypt, small hunting boats were used but they also built vessels of considerable size on the Nile, apparently propelled by many oars and guided by a large steering-oar. Sailing ships were already known by 3300 BC. Perhaps the earliest representations of boats come from the decoration on pre-dynastic Egyptian pottery of about 3300-3100 BC. Typical examples are illustrated in Figure 5.1. The top figure shows two multi-oared boats with three steering oars, miscellaneous deck fittings and 52 or 54 oars. The middle figure shows a boat of 76 oars and, near the bows is a rectangular structure on a mast or post that Bowen (1960, 117-31) takes to be the earliest representation of a sail.

However, the first definite sailboat, and one without oars, is illustrated in Figure 5.1 (bottom). It has a proper stem and stern and the mast is fitted with a rectangular sail. Bowen believes all Egyptian sails were held between a yard and a boom, largely because they developed from an identification banner fixed to a mast, but others think that the boom was a later discovery.

Shipbuilding, probably, was practised in almost every town along the Nile. Figure 5.2 illustrates the frantic activity at an Old

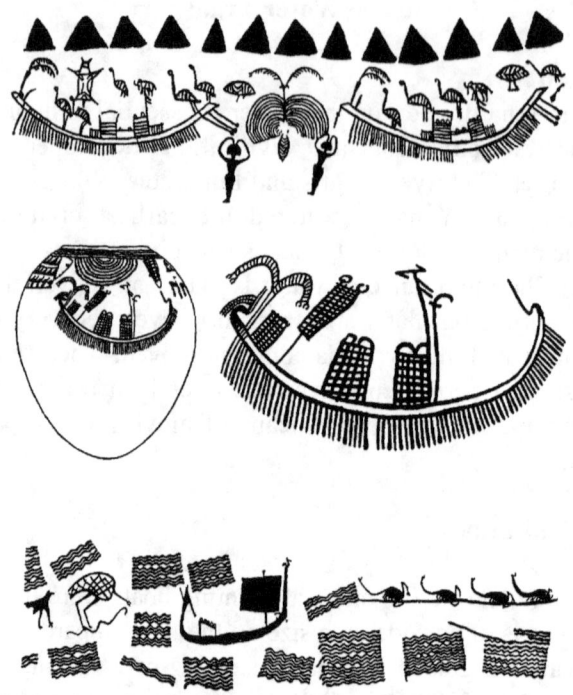

Figure 5.1 Early boats painted on Egyptian decorated Gerzean period vases.
Top: multi-oared boats with three steering oars. Ashmolean Museum.
Middle: a multi-oared boat with a primitive sail, about 3200 BC, British Museum.
Bottom: the oldest definite representation of a sailing boat, about 3100 BC.

Kingdom boat yard. There were many types of craft: the heavy cargo boat for grain and cattle, the many-oared "dahabiyeh" of the nobles, with its huge sail, and the magnificent royal burial ships. During the reign of Snefru (2613-2589 BC), for example, vessels 21 m long were built. Snefru opened-up commerce with the north and sent a fleet of forty ships to the Phoenician coast to obtain cedar logs, required for shipbuilding and other purposes. This is, perhaps, the earliest recorded example of maritime trade. We need not doubt that even larger ships were built. Snefru's son, Khufu (Cheops) for whom one the great pyramids at Giza was built, was carried to his tomb on an imposing ship

Figure 5.2 Shipbuilding in the Old Kingdom. Hieroglyphics are removed.

that has survived in almost perfect condition when released from a sealed 31 m rock-cut pit. At 43 m, it was too long for the pit and was stored as 1224 pieces. After many years of patient restoration, the ship was finally put on display in March 1982. A second sealed pit has been discovered but has not yet been excavated.

Much of the granite needed for the extensive building work of successive pharaohs was obtained from quarries at the first cataract on the Nile. From here, it was shipped downstream on boats. The site needed to be protected from the Nubians to the south but once this was accomplished Merenre (2283-2278 BC) commissioned an unbroken water passage to the granite quarries by opening a succession of five canals through the granite barriers of the cataract. His engineer Uni completed this arduous task within a year, besides building seven boats, to carry great blocks of granite for the royal pyramid. The first cataract was now passable for Nile boats during the annual period of high water and opened-up northern Nubia for increased trade and for eventual conquest. Commerce developed with lands to the south and with the land of Punt (probably northern Somalia or Djibouti). As there was no connection by water between the Red Sea and the Nile, a caravan route was established between harbours such as Kosêr or Leucos Limên on the Red Sea coast and Coptos on the Nile.

Queen Hatshepsut [1498-1483 BC] seems to have been relatively untroubled by foreign enemies. The walls of her tomb record that she promoted a great trading expedition to Punt. A fleet of five sea-going vessels carried merchandise for barter and a great stone statue of the queen, to be erected in Punt. The tomb relief, Figure 5.3 shows the

Figure 5.3. The expedition to Punt.

ships being loaded in Punt. Amongst the cargo were fragrant woods, myrrh-resin, fresh myrrh-trees, ebony, ivory, green gold of Emu, cinnamon wood, incense, eye-cosmetic, baboons, monkeys, dogs, skins of the southern panther, and natives with their children.

Despite the passage of a thousand years, Hatshepsut's ships had the same hull shape as those on earlier wall-reliefs, but the lines were more graceful and the ship looks stronger. For the first time a keel is used but it is not known if the hull was supported by ribs or by rope as on earlier boats. There was a raised foredeck and afterdeck and the ship was driven by sails and by fifteen rowers. The stem post was straight and the sternpost was curved and decorated with a lotus bud. The decking beams had been extended through the hull to increase its strength and enabled them to eliminate the rope netting that was formerly used. A rope truss on forked supports remained from earlier days but low, wide sails set amidships on a pole mast had replaced the tall narrow sails of the earlier craft. The mast was shorter and required fewer stays but the wide sails needed numerous lifts to support the boom. Its position amidships suggests that it could sail a point or two into the wind and gave the crew confidence that they could return to their homeport. A massive steering oar on the starboard side replaced the steering oars on each quarter. Landström (1969, 18), in his reconstruction, gives the Punt ships 15 pairs of oars, a hull length of 25 m, a beam of 5 m, a draught of about 1.3 m, and a sail area of 130 m^2.

Heredotus (2.96), writing in the 5th century BC, describes the construction of Nile cargo-boats from acacia wood. "They cut short planks, about three feet [one metre] long from this tree, and the method of construction is to lay them together like bricks and through-fasten them with dowels set close together, and then, when the hull is complete, to lay the benches across the top. The boats have no ribs and are caulked from the inside with papyrus. They are given a single steering oar, which is driven down through the keel; the masts are of acacia wood, the sails of papyrus." He tells us "these boats cannot sail up the river without a good leading wind, but have to be towed from the banks".

One of the greatest trading nations to emerge from the upheavals at the end of the 2nd millennium BC were the Phoenicians who occupied the coastal strip of what is now the Lebanon and northern Israel. Cedars and junipers were their chief exports and they had expertise in shipbuilding. From the 8th century BC Phoenician colonies spread along the southern shore of the Mediterranean Sea, the most notable being Carthage, which was to challenge Rome itself. Around 600 BC they were employed, by the Egyptian Pharaoh Necho and charged with the circumnavigation of Africa. Herodotus (4:42) says that a fleet manned by a Phoenician crew sailed from the Red Sea, round Africa, and returned to Egypt and the Mediterranean by way of the Pillars of Hercules [Straits of Gibraltar]. In the third year, they returned and reported that as they sailed on a westerly course round the southern end of Libya [Cape of Good Hope] they had the sun on their right – to their north (they had crossed the equator). Some take this as evidence that the voyage really took place while others are more sceptical.

In the first half of the first millennium BC warships with two-banks of rowers (biremes) and a pointed ram at the bow first emerged, see Figure 5.4. With two banks of rowers, these ships could develop greater speed, which enabled them to manoeuvre into the best position to ram an enemy. The Greeks of this time also illustrated such ships on their pottery but until the 5th century BC the Phoenicians were the better sailors, and the likely originators. Later their power declined and supremacy in the Mediterranean passed successively to the Greeks, the Etruscans and the Romans.

Figure 5.4 Phoenician warships, with sails and pointed prows for ramming, and cargo ships, all with two-banks of rowers, 700 BC. (British Museum)

Alexander the Great captured the Phoenician cities of the Lebanon in the 4th century BC and eventually they were absorbed into the Roman Empire as part of Syria. Their most famous contribution to civilisation was of course the alphabet we now use.

Greece and Rome

As Figure 5.4 illustrates, a distinct difference had emerged between warships and merchant vessels. With the invention of the ram, a new and fearsome weapon, namely the ship itself, came into use. The Greeks fully exploited this weapon and in the 6th and 5th centuries BC soldiers on ships had little importance and were kept to a minimum. Ramming contests between ships and clever manoeuvres came to be accepted practice. Skill in rowing and steering, and a good turn of speed using many rowers, was all-important. Sails were furled before going into battle and hulls were narrow and long, for lightness, to accommodate the rowers and to increase speed.

Warships were constructed of a lighter wood than merchantmen (fir) and because it was subject to rot and attack from marine borers, they were hauled from the water at every opportunity. In their homeports, they had special buildings to house them but elsewhere they were beached. There was no spare room for the crew when they were not at their oars and consequently they put in to shore at night and even at their midday meal. Accordingly, they had to travel near to land and crossing the open sea was avoided although quite common. The summer months were the sailing season.

By the 5th and 4th centuries, the standard warship was the trireme, which used three sets of oars per side, but exactly how they were arranged is still debated. It was the ship used most famously at the battle of Salamis, in 480 BC, by both the Greeks and the Persians when Xerxes invaded Greece.

Athenian inscriptions show that the triremes carried as many as 170 rowers and there were 30 other officers, crew, and marines. Heredotus (7:185) states that the Greeks of Thrace and the islands supplied 120 ships and 24,000 men to defend Greece against Xerxes, thus he allowed 200 men per ship. Landström (1969, 47) estimates the dimension of a trireme with 65 pairs of oars as 41 m length, 4 m beam (5.5 m with outrigger), and a sail area of 150 m^2. Diodorus (XIV, 41,3-42, 2), says quadiremes and quinquiremes were developed at the beginning of the fourth century B.C.

Merchant ships, as opposed to warships, required a strong and capacious hull for cargo and consequently were broad in comparison to their length. They could be rowed, but because rowers put up costs their numbers were restricted. Sails were preferred, when the wind was favourable. In extreme form, merchant ships were so broad in the beam that they relied wholly on sail and could not be rowed effectively. Aristotle thought such ships, under oars, were like large insects with inadequate wings, trying to fly. The contrast between warships and merchantmen is apparent in Figure 5.5, which shows a 6th century BC pirate ship, using both sail and oars, chasing and ramming a much larger cargo vessel, also under sail.

Figure 5.5 A fast warship, or pirate ship, with a ram attacks a large, heavy merchantman. 6th century BC.

Two merchant ships of this period have been found and excavated and their cargoes partially recovered; see Bass (1972, 50-52). About half the hull of a 4th century ship has been excavated off Kyrenia, Cyprus, and a more fragmentary 5th century BC wreck has been excavated at the Straits of Messina. Bronze coins and carbon 14 dating of the cargo (almonds) and of the Kyrenia ship's planking suggest she was built in 389±44 BC and was at least 80 years old when she sank. Both these ships were built of pine and both had their wetted surface covered in lead sheeting to protect against marine borers. Although this made them very heavy, they could be left in the water for extended periods and they could be heavy because they did not require constant hauling from the water. On the Kyrenia ship were lead rolls and a mallet, presumably for repairs to the lead sheathing. They were deep, wide vessels, that relied mostly, if not entirely, on sail. The mast position was well forward which suggests the ship carried a lateen sail that would enable her to sail into the wind rather better. These two wrecks have shown that lead sheathing, well known on later Roman ships, and lateen sails, which were thought to have been used from the 2nd century BC, were both used as early as the 5th century BC.

The Kyrenia vessel is estimated to be 25 m long and both ships were built by the well-known method in which the shell is built first and the frames are inserted second. The strakes are held together, edge to edge, by an elaborate system of mortise and tenons fastened with wooden pegs. The frames are fixed to the shell from the outside using copper nails. For steering of both warships and merchant ships the crew continued to depend on a pair of steering oars, and it was not until the 13th century AD that a single, central rudder came into use.

Glotz (1965, 294-5) and Casson (1972, 287-9) have collected many data from classical literature on the speed of voyages when sailing under favourable or unfavourable wind conditions. These data are shown in Figure 5.6. The average speed attainable in favourable winds was about 6.8 knots, based on a sailing day of 16 hours (about 200 km per day) and about 3 knots with an unfavourable wind. The latter is remarkably slow, only walking pace.

Figure 5.7 suggests that the usual practice in the Mediterranean was to build the shell first, and then to stiffen it with internal ribs; the reverse of modern practice. A Roman shipbuilder is at work shaping a rib before inserting it into a hull that is already complete. A Roman ship of the second century is shown in Figure 5.7 (right), with lateen

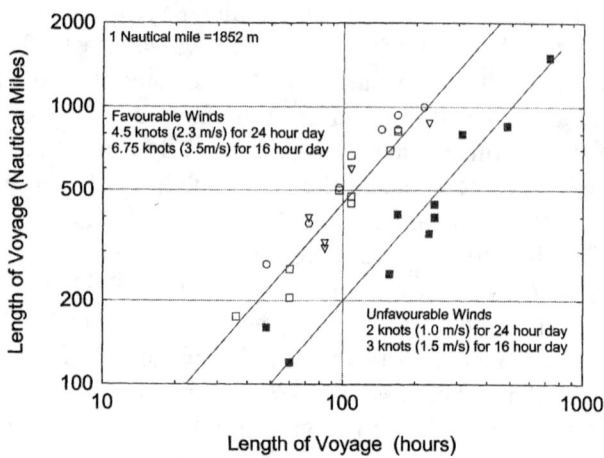

Figure 5.6 Distance sailed, and approximate time, in favourable and unfavourable winds, assuming 16 hours sailing per day.

Figure 5.7 Left: a Roman shipbuilder completes a rib before inserting it into a completed hull. This is the "shell first" method of construction. Late 2nd or early 3rd century AD. Tomb of Longidieni family in Ravenna.
Right: a Roman ship of the 2nd century with a lateen sail.

sail. Such sails enabled the ship to sail 30-45 degrees into the wind, and were commonly adopted.

The sophisticated boats of the Mediterranean may be contrasted with the simple, skin-covered boats that the Romans found in northern Europe. Hornell (1946, 112) notes that Julius Caesar, fighting in Spain in 49 BC made the first written record of British boats. Opposing Pompey's forces, Caesar had his lines of communication cut by floods that carried off two bridges built across the River Segre. By building boats like those he had seen in Britain, Caesar managed to retrieve the situation. "First the keels and ribs were made of light timber, then the rest of the hull of the boats was wrought with wickerwork, and covered over with hides." A century later Pliny, in his Natural History (4.30) also refers to British skin covered boats. He writes "Timaeus the historian says that an island called Mictis [Isle of Wight] is within six days' sail of Britannia, in which white lead [tin] is found, and that the Britons sail over to it in boats of osier, covered with sewed hides." Later he says, "At the present day they [the boats] are made in the British Ocean, of wickerwork covered in hides". Six days may have been time taken by hide covered boats, filled with tin, on a voyage from the Cornish mines to the place where merchants from Gaul took delivery of their cargoes.

Medieval and Renaissance

The oldest images of northern ships from Norway, Denmark, and Sweden, are on rock-carvings and picture-stones, Figure 5.8. They show primitive boats of the type first seen in early Egyptian illustrations; they are of about 1000 BC, and are usually described as Bronze Age, see Kühn (1956, 168-206). Many Bronze Age features survived into the early medieval period. For example, the projections beyond the bow and stern of the boat shown in Figure 5.8 are thought to be carrying handles for transporting the boat overland and are like those found on the 3rd century Hjortspring boat. The latter is 13 m long and 2 m wide. It is clinker built with five overlapping planks, each 51 cm wide and 1.5 cm thick. Evidently, an independent development had occurred in northern Europe and becomes important during the early medieval period. The planking on northern hulls overlapped (clinker

Figure 5.8 Bronze Age rock engraving of boats and men, Hogdal, Sweden.

built) rather than meeting edge to edge (carvel built). Even though clinker built hulls were more prone to leakage than carvel built hulls, this northern practice persisted through the middle ages and it was not until the early modern period that it declined in popularity.

Roman authors imply that sails were known in northern Europe, but it may be that sails were not generally adopted in England and Scandinavia until the 6th or 7th century AD. In the 7th and 8th centuries, images on pictorial stones show round-ended vessels with a single mast, a square sail, a side rudder, a proper keel, and cutwaters fitted to reduce leeway when sailing into the wind. Landström (1969, 63) reports that modern copies of such ships have a great ability to sail even to windward. They were an order of magnitude better than their predecessors and correspond to the great expansion of the Vikings out of Scandinavia, which started in about 800 AD. For example, a ship found at Oseburg was built of oak in the 9th century. It is 21.4 m long 5.1 m wide, with 12 planks each side of its keel. As with all clinker-built ships the hull was built first and the ribs inserted afterwards. The ship could be rowed by 15 pairs of oars or sailed, and she was fitted with the usual steering oar. The mast was held in a stock above the keel and supported over four deck beams.

Another famous example of a Viking longboat is that found at Gokstad Farm in 1880, under a burial mound, Figure 5.9. Her overall dimensions are 23 m length, 5 m beam, 12.2 m mast. Bows, stern and keel are made from a solid block of timber and she has 16 strakes of clinker planking tied to the ribs by means of cleats. The hull is pierced for 16 pairs of oars and when unearthed she had 32 overlapping decorative shields on each side. Although she was a burial ship, and perhaps only an elaborate model that never put to sea, she was nevertheless seaworthy. A replica sailed across the Atlantic in 1992 to be shown at the Chicago World Fair held to commemorate the 500th anniversary of Columbus's discovery of America, and to remind us that the Vikings had made this discovery almost 500 years before Columbus.

The Viking long boats were copied by others and remained in use until the 12th century before any substantial changes occurred. They were used during the Norman invasion of England in 1066. The Bayeux tapestry, which records the invasion in considerable detail,

Figure 5.9 Exterior views of the Gokstad Viking ship, 10th century,

shows the Norman invasion fleet in preparation. Timbers are felled in the forests, planks are cut to shape, the hulls assembled and drawn down to the shore, Figure 5.10 (top). It was not a hastily constructed fleet but, if the tapestry is accurate, the hulls were brightly painted and the sails were striped in three colours. Figure 5.10 (bottom) shows the fleet under sail. The actual crossing of the channel took place in

Figure 5.10 Top: Norman longboats being built for their invasion of England.
Bottom: Norman longboats sail for England.

darkness to avoid the English fleet and they landed unopposed at about 9.00 am. The ships are shown with one large sail on a central mast, and although some ships have holes for up to 16 pairs of oars, none is shown being rowed.

During the 12th century, trade was becoming more important and cargo ships emerged that were shorter and wider than long boats. The best pictures of such ships are found on the seals of port towns, although they are usually shown too short and too high. They were not rowed, since the images show neither oars nor holes for oars, but were sailing ships. Sails were preferred for cargo ships because they required fewer crew and were, therefore, less expensive to operate. Landström (1969, 72-74) presents an excellent series of these seals to illustrate the development of medieval ships. The Hastings seal has an after castle but the Ipswich seal, Figure 5.11, shows both an after castle and a forecastle. The castles are usually shown with trumpeters and were originally temporary structures raised for defence, but they became permanent features and grew. The ship on the Dover seal has castles that extend over the ends of the ship and a peculiarity is that the rudder is shown on the port side; apparently, the engraver forgot that the seal's impression would be reversed. The rudder or steering oar was always positioned on the starboard side. The other side was called the port side because, being free of the steering oar, it was the side that was docked against the quayside when in port.

Ships were getting larger and sails higher so if the ship heeled over in a beam wind the steering oar could come out of the water leaving it rudderless. To avoid this, the rudder was moved to a central position on the sternpost. The earliest image of such a rudder is a relief carved on the font in Winchester Cathedral. The font is thought to be of Belgium manufacture from about 1180. A ship steered by a central rudder, of a slightly later date, is also illustrated on the Ipswich town seal, Figure 5.11.

It was about this time that the compass first came into use in Europe. A simple compass consisting of a magnetised needle floating on wood in a bowl of water was used both in China and Europe in the 12th century, Lane (1963, 605-617). Peter Peregrinus in 1269 wrote a treatise describing several ways of assembling magnetised needles, pivots, and ways to relate the needle to the points of the horizon and the path of the ship. He did not, however, describe the compass card or

Figure 5.11 Ipswich town seal, said to be the first depiction of a central rudder steered ship, 1200 AD. Another is on the Winchester Cathedral font.

wind rose attached to the ship so that a magnetised needle suspended above the card would indicate the ship's bearing as the ship and the card changed direction. Amalfi, about 1300, appears to have made this final step. Winter voyages out of sight of land suddenly became possible.

The cog was a type of trading ship, associated with the Hanseatic League, that emerged in the 13th and 14th centuries. They had a straight bow and stern that ran down to the keel, a high freeboard, a stern rudder, a deep draught, sharp ends, and a long lateral plane. Consequently, they sailed better than shallow draught vessels with curved stems. Landström (1969, 77) considers them the successors to the Viking cargo boats with their cutwaters. There is, however, considerable doubt over what a cog really looked like. English cogs, unlike those of Hansa, did not have straight stems and long keel but were built with the traditional curved stem that joined the keel further aft. Like earlier ships, cogs were fitted with forecastles and after-castles from which pirates and other enemies could be fought. The distinction between warships and cargo ships vanished. The largest cogs were about 240 to 300 tonnes and had a single mast although some requiring two masts are known. The beam was rarely over 9.5 m and the length was 25 to 30 m, giving them a length to beam ratio of about 3:1 and a large cargo capacity. Sail area was about 175 m^2 and with bonnets fitted in mild weather it could be as much as 335 m^2.

The caravel had three masts, two and sometimes all three being fitted with lateen sails. They were suitable for the open sea but being smaller, about 65 tonnes to 100 tonnes, they were also able to sail close to shore and up the river estuaries. The lateen sails enabled them to sail into the wind, and if there was none, they were small enough to be rowed. This made the caravel the favourite ship for the great 15th and 16th century voyages of discovery, although the lateen rig was not entirely suitable for the Atlantic. Columbus, on his voyage of discovery in 1492 paused in the Canary Islands to change the *Nina* from a lateen to square rigging. On the other hand, it was the lateen sail that gave the Portuguese navigators the confidence to sail down the east coast of Africa, knowing they could return by tacking against the prevailing north wind.

In northern Europe ships remained clinker built with overlapping strakes, and during the 15th century they became much larger. Kemp (1978, 70) notes that a small revolution had taken place. At the beginning of the century a ship of 250 tonnes was considered large but by the end of the century 1000 tonne warships were being built and a third or even a fourth mast was necessary. Guns were instituted on ships at the beginning of the 14th century, a few years after their first use on land, and eventually they changed naval tactics and the design of ships. Ships could be disabled, and even sunk, from a distance and boarding, with its hand to hand fighting, was not strictly necessary. The first record of a European ship being sunk by gunfire is from 1513. When heavier cannon was introduced they had to be mounted nearer the waterline to avoid any tendency to overturn, and soon, heavy naval cannons, weighing perhaps 2 tonnes each, were mounted all round the ship, as Figure 5.12 shows. Warships were once more purpose built by the state. Henry VII, in his wars with France, built the *Jesus* of 1000 tonnes, the *Holigost* of 760 tonnes, and the *Trinity Royal* of 540 tonnes and numerous other large ships capable of

Figure 5.12 A 15th century warship with guns mounted on forecastle, after-castle, and through ports in the hull.

carrying more than a hundred men.

Another problem introduced by naval cannon was that, as the recoil momentum was equal to the momentum of the cannon ball, the cannon could inflict as much damage on their own ship as the cannon ball on an enemy. Ships needed to be reinforced and strengthened.

Early Modern

Scientific and inventive minds in the 17th and 18th centuries, turned to investigating the best form of ship's hull. Huygens in Paris and Newton in Cambridge knew that the resistance to motion of a body in water increased as the square of velocity. In 1669, Huygens constructed a towing tank for his experiments, and Bernoulli, who published his famous theory, relating to the pressure and velocity of fluids, in 1738, placed this on a firmer theoretical base although this quadratic relation is true only if wave-making resistance is small. Bouguer in 1746 also used a towing tank to investigate the resistance of various shapes. He determined the resistance of these shapes and compared them to the resistance of a flat plate, and from this he deduced what he considered the optimum hull form made up of simple plates. Both Huygens and Bouguer also investigated the relation between wind direction and sail angle and derived equations to optimise the propulsive force of the sails. Their theories are discussed below.

The foremost naval architect of the period was Chapman (1721-1808); see Harris (1989). He was the first to build ships on scientific principles and his publications became the standard textbooks of his profession, first in Sweden and later in France and England after they were translated. He was the son of a British naval officer who had joined the Royal Swedish Navy, where he did most of his work, and published some 15 extensive technical papers.

The French Academy of Science offered prizes for the best papers on naval architecture and this prompted the work of Huygens, Bouguer, Euler, and others. The British lagged behind continental Europe, at least in the theoretical understanding of maritime science. Brown (1990,15-24) has reassessed the British response to this problem. In 1791, The Society for the Improvement of Naval Architecture was founded at the Crown and Anchor in the Strand. Membership including the future King William IV, the First Lord of the Admiralty, the President of the Royal Society, and 270 other notable figures with maritime interests. They offered awards of up to £100 for original work on the theory of floating bodies and their resistance to motion, to obtaining plans of ships, to calculating their capacity, centre of gravity, etc, and to carrying out experimental work. The first paper was on ship stability, by an anonymous naval officer, who described his experiments, moving 14 three-ton guns through 3 ft and measuring the change in the angle of heel, from which he deduced the metacentric height. Chapman came over from Sweden to set out the full theory of these experiments in a further paper, and the Society re-published George Attwood's papers to the Royal Society in 1796 and 1798 on ship stability at large angles of heel.

Their most famous work was in sponsoring Colonel Beaufoy's experiments on resistance. Beaufoy carried out some 173 tests of model ships in a towing tank that allowed runs up to 300 ft [91.5 m] in length using models 30 to 42 ft [9-13 m] long. Data on some 1844 tests were published by the Society in 1800, 1814 and 1834. However, he did not know how to apply his extensive work to full size ships; a problem that was not solved until Froude conducted his classical trials some 70 years later. The Society was wound up in about 1799.

Improvements in the longitudinal rigidity of ships, and in the theoretical understanding of longitudinal bending were also made about this time. Robert Seppings (1767-1840), the Master Shipwright at Chatham in 1804 noted that hulls were made up of transverse ribs covered with longitudinal planks, like a farm gate without diagonal cross-members, and thus weak in shear. If a diagonal cross-member is added to a gate it is much stiffer in consequence. This was Sappings' only theoretical justification for his work and it turned out to be correct and allowed him to construct much larger and stronger wooden hulls.

The system of diagonal braces was so successful that Seppings presented papers to the Royal Society in 1814 and 1817, and was knighted. When he applied diagonal bracing to a frigate it halved the hogging deflection, when it was floated, and when applied to new ships his method required 180 fewer trees. Seppings' work was important because if hulls could be made stiffer, they could also be made longer and could support a heavier load. The restrictions on size that had held wooden ships in check for four hundred years were partially lifted.

Following Seppings' 1814 paper, Thomas Young also presented a paper to the Royal Society in which he provided theoretical support to Seppings' ideas. He also attempted to calculate a ship's deflection using a well-known method of calculating the deflection of prismatic bars from the bending moment; a method first employed by Euler in relation to the collapse of elastic columns.

The design of the sailing ships improved and ultimately a fleet of fast sailing clippers maintained the North Atlantic service, carrying mails and passengers (immigrants). The *Fidelia*, a clipper of the Black Ball fleet, ran from New York to Liverpool, in 13 days 7 hours (2855 nautical miles at 8.9 knots), and returned to New York in 17 days 6 hours (6.9 knots). She was built in New York by William Webb, who was responsible not only for many of the finest packet ships, but also for several of America's best clippers. Clippers were always considered the perfection of the ship builder's art, and their beauty was greatly admired. In the early 19th century, before the American Civil War, the New England clippers could out-sail any merchant ship. A clipper sailing from New York to California via Cape Horn was the fastest way between these places until the transcontinental railway was built. Britain's answer to the American clipper was the Blackwall frigates that were built for the route to India via the Cape of Good Hope and, of course, the famous tea-clippers that raced to get the first of the tea crop from China, Ceylon or India, to Britain.

The growth in the size of American and British clippers is illustrated in Figure 5.13, which is based on data in Campbell (1974,

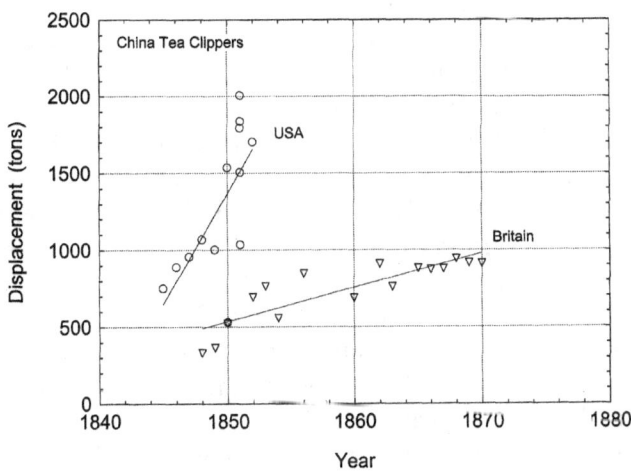

Figure 5.13 Displacement of American and British Tea-Clippers.

61). American clippers were generally about twice the size of British clippers and were usually about 10 m longer; they were also wider

135

and had a deeper draught. They carried more cargo and sailed faster than British Clippers until the advent of steamships disrupted the trade. The reason for this disparity was that American ships were built with more rigid hulls. Commercial shipbuilders in Britain were slow to adopt the naval practices devised by Seppings (diagonal bracing and filling up the spaces between the frames) whereas American shipbuilders adopted diagonal bracing much earlier.

Figure 5.14, which is adapted from Campbell (1974, 63-4), illustrates the main differences. British commercial builders kept to the old system of vertical framing spaced out, and covered, inside and out, by longitudinal planking. The wall thickness was typically about 24 inches (0.61 m) near the keel. The sealed-in timbers, if not made of teak, were an inducement to rot because air cannot circulate. This was controlled originally by charring the surfaces with hot irons and later by salting. It was found that wood resisted rot if well soaked beforehand in salt water, or packed with plain salt in the spaces left during construction. The frames of the British ships were made of two sets of timbers that abutted at the keel but gradually separated towards the extremities. This was an expensive form of construction, originally to encourage air circulation and later to be used for salting. The American system also used two sets of timbers for the frame, but these remained parallel and abutted each other continuously. The only gap in American construction was between the frames. The additional rigidity needed by the American clippers was provided by extra keel pieces and by diagonal bracing with iron straps let into the frames to provide a flush surface for the outer planking. The frames of the American ships, where they crossed the keel, were not notched, as were the British, and this removed a source of weakness.

Wooden hulls gave way to composite hulls of wood and metal, and eventually to all-metal hulls, which could be made longer. The first all-metal hull was Brunel's steamship the SS Great Britain, built in 1843, 98 m in length and propelled by sail and by steam. It is now on display in Bristol. Sailing ships, still trying to compete with steam, also adopted iron hulls.

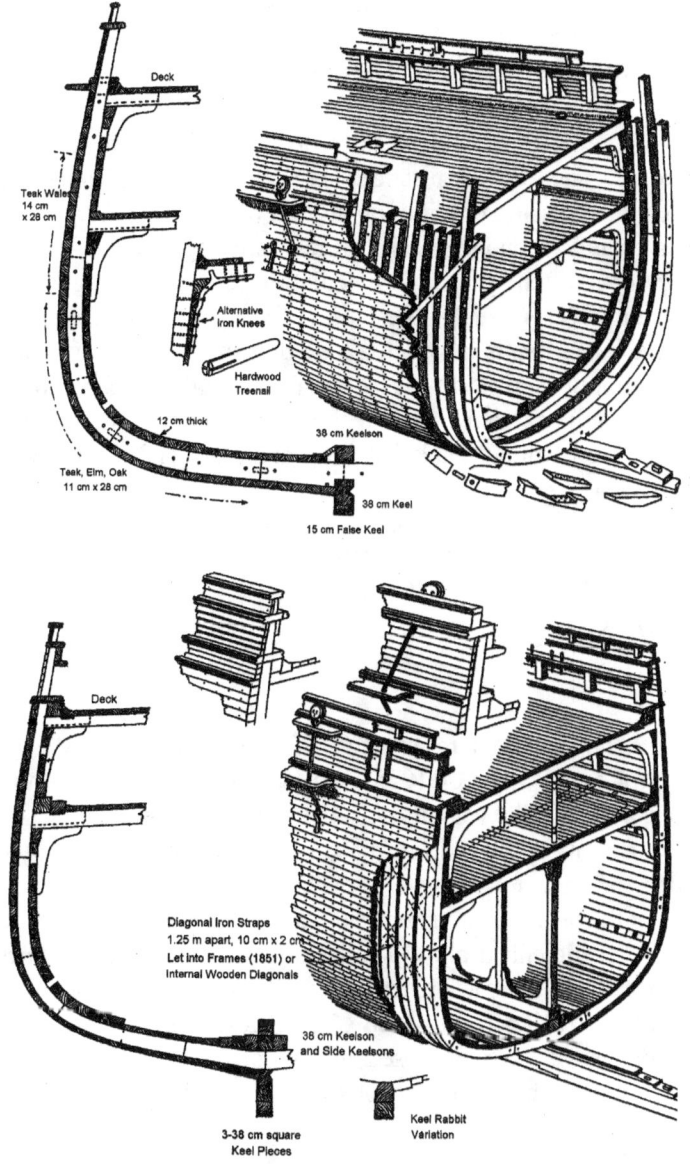

Figure 5.14 Construction of typical British (top) and American (bottom) wooden hulls for clippers. After Campbell (1974, 63-4).

Appendices

Force and Power Required by Ships

The force required to propel a ship through water must overcome: frictional resistance between the water and the wetted surface, the pressure distribution around the hull, the force to generated waves at the bow and stern, and air resistance against the superstructure. The latter is often ignored in calculations but can be significant. The frictional force is usually the largest force on the hull and is derived from the experiments in which a long thin smooth board is towed under water without significant surface wave formation. Several empirical correlations have been used but the most common is that proposed at the 1957 International Tank Towing Conference (ITTC), namely

$$C_f = \frac{F_f}{0.5 \rho_w A_w v^2} = \frac{0.075}{[\log(Re) - 2]^2}$$

$$Re = \frac{\rho_w v L}{\mu} = \frac{vL}{v_w}$$

5.1

The frictional resistance is F_f, the water density is ρ_w (1000 kg/m^3 for fresh water or 1025 kg/m^3 for sea water), A_w is the wetted surface area, v is the velocity, L is the waterline length of the ship, and μ is the viscosity. The ratio $v_w = \rho/\mu$: is the dynamic viscosity of water (1.118x10^{-6} m^2/s for seawater). The ITTC friction equation is strictly for smooth surfaces and a nominal value of 0.0004 is often added to the friction coefficient to allow for the effects of surface roughness. The wetted surface area is difficult quantity to determine and requires a precise knowledge of the hull shape However, several approximate equations have been proposed and those of Froude and Gertler are useful when the hull shape is not known in detail.

$$A_w = \nabla^{1/3} L(1 + \nabla^{1/3} / L) \quad \textit{Froude}$$

$$A_w = 2.5\sqrt{\nabla L} \quad \textit{Gertler}$$

5.2

V is the volume of water displaced by the hull and L is the waterline length of the hull.

Wave making resistance depends mainly on the Froude number $Fn=v/(gL)^{0.5}$, where v is the ship's velocity, g is local gravity, and L is the ship's length. This dependence is caused because the interference of the bow-wave with the stern-wave and occurs when

$$Fn = \frac{v}{\sqrt{gL}} = \frac{1}{\sqrt{2\pi(n-0.25)}}$$

$n = 1, 2, 3, ---$
$Fn = 0.46,\ 0.3,\ 0.24,\ ---$
5.3

The non-dimensional number Fn is called the Froude number after William Froude who carried out much of this work in the 1870s; see Froude (1874). When the Froude number is equal to one of the critical values indicated in Equation 5.3 then wave reinforcement occurs and the drag resistance is increased. Cancellation of the stern wave occurs when a crest from the bow wave corresponds to the position of the stern so that

$$Fn = \frac{v}{\sqrt{gL}} = \frac{1}{\sqrt{2\pi(n-0.75)}}$$

$n = 1, 2, 3, ---$
$Fn = 0.8,\ 0.36,\ 0.27,\ 0.22$
5.4

Equation 5.4 defines the Froude numbers at which the bow and stern waves cancel thereby reducing the drag resistance of the ship. Thus, wave making resistance does not increase steadily with increasing speed but exhibits a series of humps and hollows corresponding, approximately, to the Froude numbers at which, theoretically, the bow and stern waves reinforce or cancel. In practice, the main hump is found to be at $Fn=0.52$ (rather than 0.46), the second hump, often called the prismatic hump, occurs at about $Fn=0.31$ (rather than 0.3), and the third hump occurs at $Fn=0.26$ (rather than 0.24). The Froude number of rowed ships and sailing ships never exceeded about 0.2 and so wave reinforcement and cancelling is not important.

Experimental data show that the coefficient of resistance caused by wave making may be represented, if Froude number is not too large, by

$$C_w = \frac{F_w}{0.5\rho A v^2} = 0.8 Fn^4 = 0.8\left(\frac{v}{\sqrt{gL}}\right)^4 \qquad 5.5$$

Thus, the coefficient of resistance increases with v^4. The total resistance to motion experienced by the hull is the sum of the frictional resistance and the wave making resistance and is written as

$$C_T = C_f + C_w = \frac{F_T}{0.5\rho_w A_w v^2}$$

$$C_T = \frac{0.075}{[\log_{10}(Re) - 2]^2} + 0.8(Fn)^4 \qquad 5.6$$

A useful relation between the Reynolds number and the Froude number is

$$Re = \frac{Fn\sqrt{gL^3}}{v_w} \qquad 5.7$$

Substituting Equation 5.7 and 5.1 into Equation 5.6 gives a relation between the total drag coefficient and the Reynolds number or between total drag coefficient and Froude number.

$$C_T = \frac{0.075}{[\log_{10}(Re) - 2]^2} + 0.8\left[\frac{v_w Re}{\sqrt{gL^3}}\right]^4 \qquad 5.8$$

This relation is plotted in Figure 5.15 (top) for hulls of 10, 20, 50 and 100 m length. As the Reynolds number increases with speed the total drag coefficient decreases until at some point the wave making resistance starts to increase rapidly. For a 10 m hull this occurs when the Reynolds number is about 10^7 and for a 100-m hull it occurs when Reynolds number is about 3×10^8. In each case the wave making resistance increases rapidly when the Froude number exceeds 0.1 and usually, in ships propelled by oars or sails, the Froude number does not

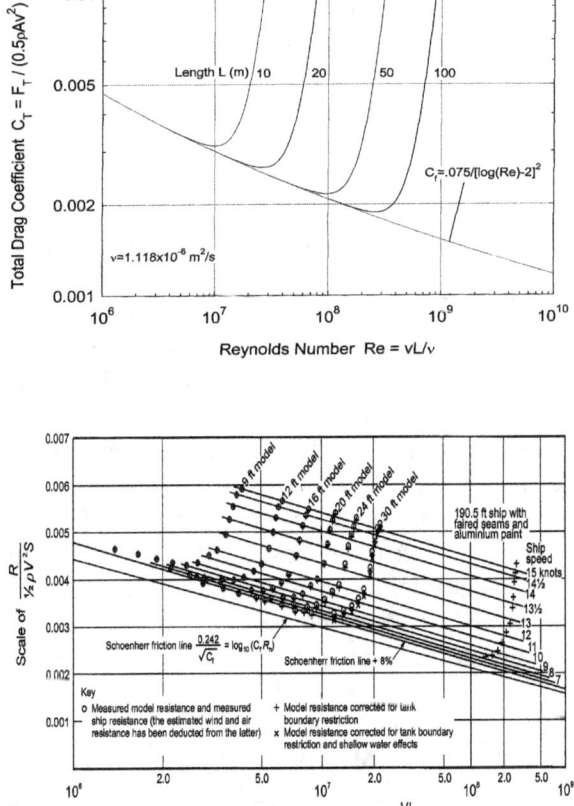

Figure 5.15 Top. Variation of ship's resistance with Reynolds number. Bottom. Measured variation in ship's resistance with Reynolds number for a full-size ship and six models of diverse sizes. After Muckle and Taylor (1987, 286)

141

exceed about 0.25. Figure 5.15 illustrates very clearly that for a given propulsive force, long ships have less resistance than short ships. It also illustrates the problem of using model ships to estimate the performance of the full-size ship. To overcome this difficulty the friction coefficient must be subtracted from the total resistance of the model ship and replaced by the frictional resistance of the full-size ship. Typical experimental results are shown in Figure 5.15 (bottom).

Another way of plotting the same data is to express it in terms of the Froude number instead of the Reynolds number by substituting Equation 5.7 into Equation 5.8 to give

$$C_T = \frac{0.075}{\left[\log_{10}\left(Fn\sqrt{gL^3}/v_w\right) - 2\right]^2} + 0.8(Fn)^4 \qquad 5.9$$

To demonstrate the validity of Equation 5.9 it may be applied to estimate the velocity of five early 20th century passenger liners, for which reliable data are available. The results are shown in Table 5.1. The average error in the estimated velocity is 2.7%, which is remarkably good considering that the only variables used in the calculation were the length, gross weight, and available power. Breadth and draught are given in the table but are not used in the calculation. Part of the reason for this accuracy is that velocity increases approximately as the fifth power of speed and so a 5% error in speed corresponds to a 25% error in power. However, so long as these equations are only used to predict a ship's speed they may be used with some confidence and they may be used to estimate the performance of boats powered by oars or by sails.

Table 5.1 Estimation of Speed using Equation 5.9.

Ship	Lucania	Mauretania	Olympic	Aquitania	Queen Mary
Date:	1893	1907	1911	1914	1936
Displacement m^3	12837	30422	46025	45240	80053
Weight tonnes	13157	31183	47176	46371	82055
Length m	189.7	240.8	269.1	274.8	310.7

Breadth m	19.8	26.8	28.0	29.6	36.0
Draught m	9.04	11.04	10.54	10.75	11.13
Power kW	20627	53712	41020	44760	117868
Speed m/s	11.3	13.5	11.3	12.1	14.7
Calculated Speed m/s *	11.3	12.85	11.8	12.3	14.1
Error % **	0.0%	4.8%	4.4%	0.2%	4.1%

* Assuming propeller efficiency = 65% **Average error = 2.7%

Strength of Hulls

As shown above, long ships, of a given displacement, experience less resistance to motion than short ships and it seems reasonable therefore for shipbuilders to show a marked preference for long ships. However, other factors such as strength and displacement are also important. Warships could be built long and slim to give them a good speed but merchant vessels need space for cargo so they were made wider in the beam.

Ships need a rigid structure to resist the longitudinal bending stresses. Stress analysis of a ship's structure is a complex and long process and there is insufficient space here to do it justice. However, a simpler approach leads to useful results and illustrates the way the hull's strength limits its length and displacement. The worst condition occurs when the centre of the hull temporarily rides on a large wave and the bow and stern are out of the water. This causes the ship to hog (droop at the bow and stern relative to the centre). The maximum bending moment is experienced near the mid-section. If the buoyancy force is concentrated at the mid-section and the weight is distributed on either side such that half the weight acts at a distance $L/3$ from the mid-section (to allow for the plan shape of the ship and the distribution of weight) then the maximum bending moment is $M=WL/6$. W is the ship's weight and L its waterline-length. From the theory of bending

$$M = \frac{WL}{6} = \sigma Z \qquad 5.10$$

σ is the allowable longitudinal stress and Z is the section modulus. For simple boats with a decked hull, the section modulus is proportional to the beam B, the hull thickness t and the section height above the keel H, and a good approximation is $Z=kBHt$, where k depends on the cross-section that is adopted. An average value might be $k=0.25$. Substituting into Equation 5.10 gives the maximum displacement of the hull:

$$W = 1.5\sigma BHt / L \qquad \qquad 5.11$$

Long narrow ships can carry less weight than short wide ships. The displacement may also be calculated from

$$W = \rho g C_B BDL \qquad \qquad 5.12$$

D is the draught, that is, the distance between the bottom of the keel and the waterline when fully laden, C_B is the block coefficient $C_B = \nabla / BDL$, ρ is the density of water, and g is local gravity. From Equations 5.11 and 5.12 the maximum length of the hull at the waterline is

$$L = \sqrt{\frac{1.5 \; \sigma \; t \; H}{\rho g C_B \; D}} \qquad \qquad 5.13$$

This length must not be exceeded if the stress is not to exceed the maximum allowable stress of the hull. For a wooden ship, the allowable stress is quite low, about 10 MPa, but for iron or steel ships, the allowable stress is much greater, and larger and longer ships can be built.

Equation 5.13 and the corresponding displacement are plotted in Figure 5.16 and illustrate the general limits to the displacement and length of wooden hulled ships; the curves show the trends rather than the exact limits. In plotting these curves, it has been assumed that the maximum allowable design stress in wood is 10 MPa, the density of

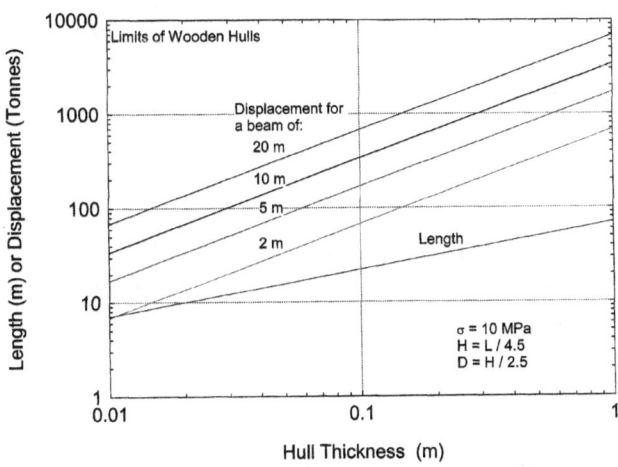

Figure 5.16 The maximum length and displacement of wooden ships having specified hull thickness and strength (approximate).

seawater is 1025 kg/m^3, local gravity is 9.81 m^2/s, and the block coefficient is 0.75. The depth at the mid-section is taken as the waterline length divided by 4.5, and the draught is taken as the mid-section depth divided by 2.5.

The resulting curves illustrate the trends in wooden hulled boats quite well. Boats having a hull thickness of about 2 cm are limited to a waterline length of about 10 m. With a beam of 2 m the maximum displacement is about 15 tonnes, if it is decked, but considerable less if it is hollow (without a deck). Increasing the beam to 5 m increases the allowable displacement to about 70 tonnes. If the hull thickness is 10 cm then the maximum waterline length is about 20 m and the maximum displacement is about 160 tonnes for a beam of 5 m and about 320 tonnes if the beam is 10 m. If the hull is 0.5 m thick then the maximum length is about 50 m and the maximum displacement is about 1600 tonnes if the beam is 10 m and 3200 tonnes if the beam is 20 m. Wooden ships of this size were rarely built. For a hull thickness of 1 m the maximum length is about 70 m and the maximum displacement is about 7000 tonnes for a 20 m beam. The largest wooden ship ever to be built was the American clipper *Great Republic*, which was launched in 1853. She was 99 m long and 16.2 m wide. However, she never sailed in this configuration for a fire broke out as she was about to start her maiden voyage and her hull had to be reduced during repairs.

Rowing

As shown above the total drag force on a boat travelling through still water is caused by frictional drag on the wetted surface and wave making drag, thus

$$F_T = C_T \rho_w A_w v^2 / 2 \qquad \qquad 5.14$$

C_T is the total drag coefficient defined by Equation 5.8 or 5.9, ρ_w is the density of water, A_w is the wetted area of the boat, and v is its velocity through the water. The wetted area may be estimated from the volume water displaced by the boat, ∇, using either Froude's or Gertler's approximations, Equation 5.2.

The drag coefficient of an oar is C_d and the velocity of the oar through the water is $\omega b - v$, where ω is the rotational velocity of the oar during the power stroke, b is the distance between the rowlock and the blade, and v is the boat's velocity. The notation is defined in Figure 5.17. If there are n oars then the force propelling the boat through still water is

$$F_0 = \alpha n C_d \rho_w A_0 (\omega b - v)^2 / 2 \qquad \omega b > v \qquad 5.15$$

α is the proportion of time required for the power stroke (typically about 0.6), ρ_w is the density of water and A_0 is the area of the oar-blade. If the blade travels at the same velocity as the boat ($\omega b = v$) then there is no force exerted by the oars on the boat; if $\omega b < v$ then the oars retard the boat, as when the oars are stationary relative to the boat. The normal situation is $\omega b > v$ so the oars propel the boat forward. This force is exerted during the backstroke and during the return stroke the

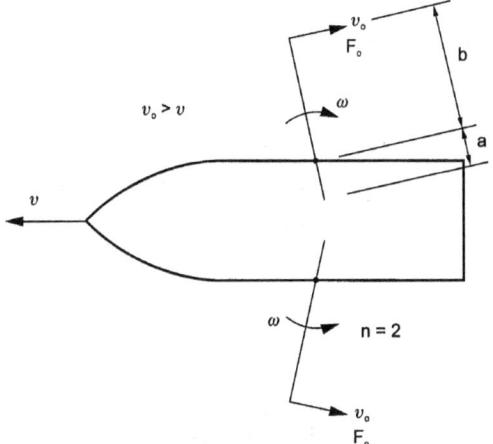

Figure 5.17 Rowing geometry and notation

force is zero. The force F_0 is the average force exerted during one complete stroke if $\alpha = 0.6$.

For the steady state, the force driving the boat forward is equal to the boat's drag through the water, that is, $F=F_0$. Substituting from Equation 5.14 and 5.15, and writing the steady state velocity as v_∞ we obtain the relation between the boat's steady state velocity and the oar blade velocity in the form

$$\frac{v_\infty}{\omega b} = \frac{1}{1+\sqrt{\Phi}}$$
$$\Phi = \frac{C_T}{\alpha n C_d} \frac{A_w}{A_0} \qquad 5.16$$

The term under the square root may be called the boat to oar drag ratio and, as will be shown, is fundamental in defining the rowing efficiency. If the positive value of the square root is taken then the boat velocity will always be less than the oar velocity, $v < \omega b$.

Now, the power exerted by the oarsmen is

$$W_0 = F_0 \omega b \qquad 5.17$$

The power required to propel the boat is

$$W = Fv \qquad 5.18$$

The efficiency of rowing may be taken as the ratio of these powers, and for the steady state $(F=F_0)$ we have

$$\eta = \frac{W}{W_0} = \frac{v}{\omega b} \qquad 5.19$$

The efficiency, η, is the proportion of the rowers' power that is expended in propelling the boat; the remainder $(1-\eta)$ is expended in creating waves and eddies by the oars. For example, if the boat is rowed into a solid obstacle so its velocity is zero then the efficiency is zero and all the power exerted by the rowers is expended in eddies and waves, not by the boat but by the oars.

For the steady state, $F = F_0$ and thus

$$\eta = \frac{W}{W_0} = \frac{v_\infty}{\omega b} = \frac{1}{1+\sqrt{\Phi}}$$

$$\Phi = \frac{C_T}{\alpha n C_d} \frac{A_w}{A_0} \qquad 5.20$$

If the total area of the oars, nA_0, is small then the rowing efficiency is small and most of the effort is wasted. If, on the other hand, the total area of the oars is large then the rowing efficiency approaches 100% although, realistically, it seldom exceeds about 80%. Every additional rower increased the power available to drive the boat and increased the rowing efficiency of the whole company of rowers. If the power developed by each rower is w_0, then the boat's power and steady state velocity is given by

$$W = \eta n w = \frac{n w_0}{1 + \sqrt{C_t A_w / (\alpha n C_d A_0)}} = \frac{C_T \rho_w A_w v_\infty^2}{2}$$

$$v_\infty = \left(\frac{2 n w_0}{C_T \rho_w A_w \left[1 + \sqrt{C_t A_w / (\alpha n C_d A_0)}\right]} \right)^{1/3} \qquad 5.21$$

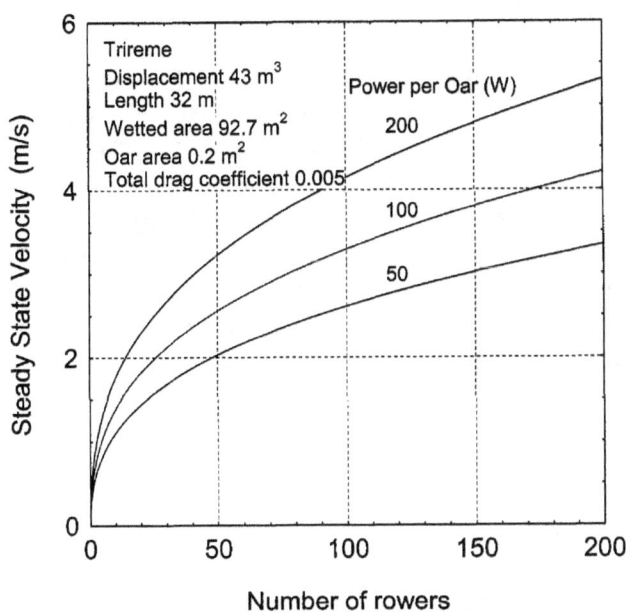

Figure 5.18 Influence of number of rowers on the steady speed of a trireme.

This equation is reasonably simple but unfortunately the total drag coefficient depends on the velocity so an iteration process is necessary to obtain a solution. However, a reasonable drag coefficient is about 0.005 and the steady velocity is then determined. Figure 5.18 illustrates the relation between velocity and the number of rowers for a trireme of 43 m³ displacement, 32 m length. The area of each oar is assumed 0.2 m², and the power developed by each rower is taken to be 50, 100 or 200 W. The wetted area from Equation 5.2 is 92.7 m², the density of seawater is 1025 kg/m³ and α=0.6. For short periods, when the ship is used as a ram, the rowers may develop 200 W per man and with 170 rowers a velocity of about 5 m/s (9.7 knots) might be attained. For extended periods of fast sailing the power developed may be 100 W/man and the speed drops to about 4 m/s (7.7 knots). For relatively easy rowing a power rating of 50 W/man would give a comfortable cruising speed of about 3 m/s (5.8 knots).

Sailing

Figure 5.19 depicts the forces acting on a sail. A wind of velocity w relative to the ship impinges on the sail with an angle of attack α. Relative to the keel the angle of the wind is γ. The sail is set at an angle θ to the keel and so $\gamma = \alpha + \theta$. The force, F, normal to the sail and has components F_x along the keel and F_y normal to the keel, see Hoerner (1965). These components push the ship forward along the keel with velocity v_x and to leeward, normal to the keel, with velocity v_y. The resultant velocity, v, is in direction φ relative to the keel. Thus, the direction of sailing is at some angle β relative to the wind where $\beta = \alpha + \theta + \varphi$. For present purposes, the leeward velocity may be ignored ($\varphi = 0$). The force generated by the wind normal to the sail is:

$$F = 0.5 C_d \rho A w^2 \sin^2 \alpha \qquad 5.22$$

C_d is the drag coefficient of the sail, ρ is the density of air, A is the area of sail, and w is the wind velocity relative to the ship. The x-component of this force is responsible for moving the ship through the water with velocity v_x and substituting $\alpha = \gamma - \theta$ we may write

$$F_x = F \sin \theta = 0.5 C_d \rho A w^2 \sin^2 (\gamma - \theta) \sin \theta \qquad 5.23$$

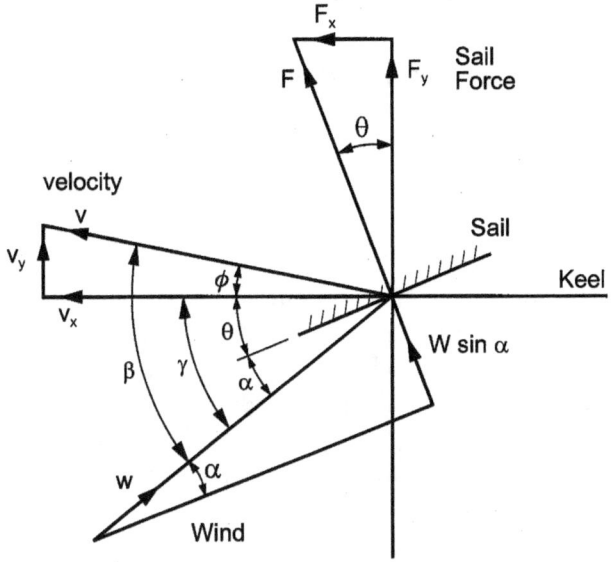

Figure 5.19 Forces generated by a sail.

Differentiating with respect to θ and equating to zero gives the condition for maximum force, F_x, namely

$$2\theta = \tan(\gamma - \theta)$$
$$\lambda \approx 3\theta$$
  5.24

This relation was first derived by Bouguer in 1757 and is known as Bouguer's theorem. Huygens in 1698 obtained a similar result but used a more complex relation.

A more modern approach is to treat the sail as an aerofoil in which the lift coefficient is directly proportional to the angle of attack ($C_L = c\alpha$). The component of the lift force in the direction of the keel is then

$$F_x = 0.5c\alpha\rho A w^2 \sin\theta = 0.5c\rho A w^2 (\gamma - \theta)\sin\theta \qquad 5.25$$

Differentiating with respect to θ and equating to zero to give the maximum force yields

$$\begin{aligned}\gamma - \theta &= \tan\theta \\ \gamma &\approx 2\theta\end{aligned} \qquad 5.26$$

This result accords with modern sailing practice for square rigged ships, see Harland and Myers (1984, 62), which is to trim the sails to bisect the angle between the keel and the wind. Of course, both these results maximise the force along the keel rather than in the direction of motion, but the resulting error is small if the drift of the ship is small. The optimum sail angle is illustrated in a painting of HMS Victory sailing into the wind, Figure 5.20.

Figure 5.20 *HMS Victory* sailing into the wind at the battle of Trafalgar, 1805 (Monamy Swaine).

Chapter 6 Undersea and Aerial Transport

Introduction

The idea of travelling under the sea or through the air has always excited the human imagination, but such ideas were mostly restricted to the realm of myth and fable. The important principle that "Any floating object displaces its own weight of fluid", was discovered at a very early date by Archimedes (c287-212 BC) and set out in his treatise *On Floating Bodies*. It is the principal on which submarines and air balloons rely, so it was quite reasonable to expect substantial progress on such devices. What attempts were made were usually related to travelling through a fluid rather than simply floating in it, and before industrialisation and the availability of suitable prime movers, such movement was difficult if not impossible. Nevertheless, some advances did occur and this chapter reviews the development of underwater craft and air balloons to the point where powered movement becomes possible.

Undersea Craft

Undersea craft such as diving bells and submarines developed quite early. The first recorded reference to human underwater activities is credited to the pseudo-Aristotlian author of "Mechanical Problems" where the "tubes used by divers" are compared to an elephant's trunk. It is said that Aristotle's pupil, Alexander the Great, descended into the Bosporus in a large barrel to observe the marine life. Alexander Neckham (1157-1217) has an illustration of Alexander the Great inside a giant glass vessel with the interior lit by suspended tallow lamps, and fantastical marine creatures surround the vessel. A little later Roger Bacon (1214-97) mentions "machines for walking in the sea, even to the bottom, without danger to life or limb".

The military advantage of being able to sink under the waves and reappear later is considerable and the commercial possibility of being able to descend to a wreck to rescue valuables is also very great. Leonardo da Vinci sketched many divers' suits, some of which had provision to supply air for breathing while submerged. The example illustrated in Figure 6.1 is of a diver and a boring machine with which he was to bore through the hull of an enemy ship without being seen.

Figure 6.1 Leonardo's diver and boring machine to sink an enemy ship.

The diver is clad in a voluminous suit, which evidently contains sufficient air for a short, mission but is not otherwise supplied with breathing facilities. It seems rather fanciful.

Kemp (1988, 192) says that William Bourne, writing in 1578 was the first to appreciate the principle on which a submarine operates and to describe how such a ship could be constructed. It is not known if Bourne's submersible ship was ever attempted, but interest in underwater devices was developing. Francis Bacon described a diving bell in 1620. It was a "sort of metal barrel lowered into the water open end downwards". This sounds very like a diving bell seen about 40 years later by John Evelyn (*Diaries, 1661*); he writes "We tried our Diving-Bell or engine, in the dock, at Deptford, in which our curator continued half an hour under water; it was made of cast lead, let down with a strong cable." The experiments at Deptford paid dividends a few years later, in 1665, when a diving bell was used to recover guns from the sunken Armada wreck *Florencia*. Two years before this, Hans Albrecht von Treileben and Anders Peckel had used a simple diving bell to salvage most of the guns from the Swedish warship *Vasa* in 1663-4. The 1660s seem to have been the time when diving bells became useful rather than interesting.

Klemm (1959, 182) says that between 1680 and 1685 Borelli devoted himself to applying the principles of mechanics to the movement of organisms and from these ideas he devised an apparatus to maintain a man under water, Figure 6.2. The man is in a watertight suit and has a large helmet from which he draws air but which is not otherwise supplied. He is weighted with lead until the top of the helmet is just level with the surface of the water. A plunger-pump *RS* is attached to his waist and as the plunger is depressed the volume of air is compressed, and the buoyancy force decreases so the man begins to sink. By adjusting the piston, he can maintain himself at any desired depth and is free to swim, or walk on the bottom, as he desires. Borelli went on to apply this principle to a closed container adapted as a boat, Figure 6.2. Inside the boat several goatskin bags, *ON*, are connected to holes *O* in the floor of the boat. As water is allowed into or pumped out of these bags the boat's specific volume changes. If the boat's specific volume is greater than that of water then the boat rises and if less it sinks.

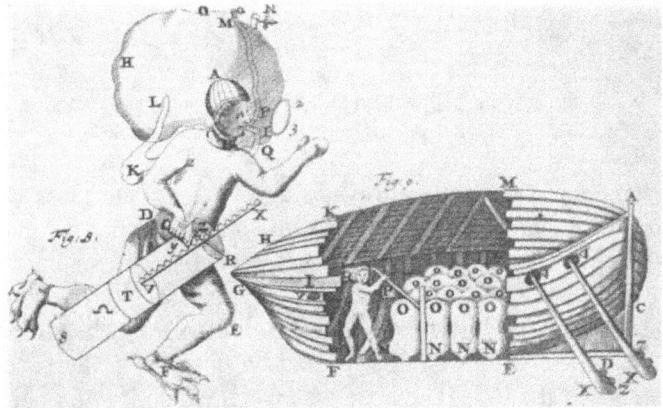

Figure 6.2 Proposed diving apparatus and submarine boat, 1680-5, engraving from G A Borelli, *De motu animalium*.

Although the idea is sound, the proposal to use a man with a pole to force the water out of the bags leaves something to be desired. A series of plunger pumps, instead of the goatskin bags, would be preferable.

Edmund Halley, the astronomer, developed a diving bell with a system for refreshing the air inside the bell and a way of providing air to men working outside the bell, Figure 6.3. Halley's bell was made of

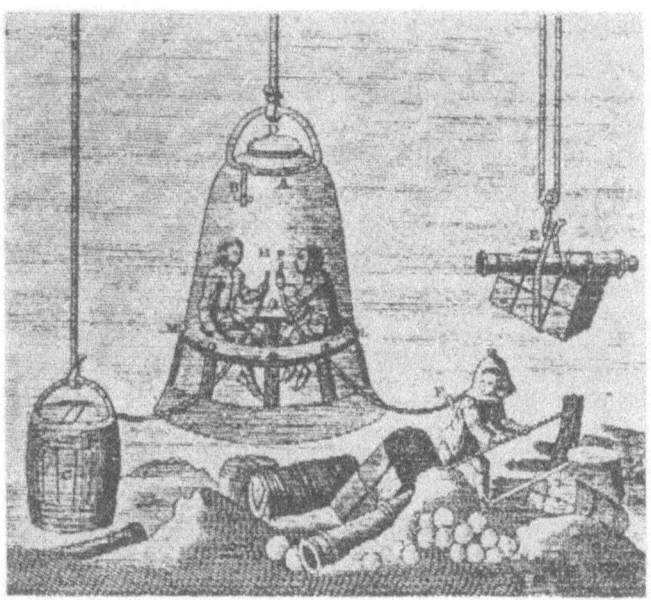

Figure 6.3 Halley's diving-bell with an air barrel supplying air by water pressure to the bell, and a diver (right) in a "cap of maintenance".

wood, weighted with lead, and was fitted with a plate glass window, and a bench for the divers. It was quite small, having a capacity of only 60 cubic feet [1.7 m^3]. It was lowered to the seabed from a mast on the parent vessel. Spent air was released by opening a tap at the top of the bell and fresh air was supplied from the surface in a barrel encased in lead. The barrel had an opening at the base and a flexible hose at the top. The hose was weighted so it remained below the bottom of the barrel and thus the air could not escape. When the barrel reached the bell, the weighted hose was drawn inside the bell and when its open end was higher than the bottom of the barrel, water entered the barrel and pushed the air from the barrel into the bell. The system seems to have worked very well. Halley says that "I myself have been One of Five who have been together at the Bottom, in nine or ten fathoms Water, for above an Hour and a half at a time, without any sort of ill consequence." Divers working outside the bell were supplied with air using a "cap of maintenance". On the death of his father in 1684 Halley also lost his financial support and was forced to give up his position and restore his family's fortune. The company he founded to develop and use his diving bell appears to have been related to this crisis. His interest in diving bells remained, however, and as late as 1716 he demonstrated a more advanced system on the Thames.

The fathers of the modern submarine are usually considered to be two Americans, David Bushnell and Robert Fulton, see Kemp (1978, 192-5). Bushnell developed his submarine, Figure 6.4 as a weapon to break British sea power during the War of Independence. It was an egg shaped, iron hull, weighted with a lead keel of 319 kg, of which 91 kg was detachable in an emergency. Two small tanks in the bottom of the submarine could be filled with seawater by a foot-operated valve and could be ejected by two hand-operated pumps. Vertical and horizontal screws could be turned by hand to provide movement and it was steered by a rudder. The *Turtle* was built and was used to attack *HMS Eagle* in 1776. The idea was to sink just under the waves, approach the enemy, then sink under their hull where 68 kg of gunpowder could be attached by means of a screw mechanism operated from inside the submarine. The attack would probably have been successful but it was not realised that the hulls of British ships were covered with copper sheets as a protection against sea worm and consequently the explosive could not be attached. The *Turtle* escaped but in two later attacks on other warships she was observed and consequently failed.

A similar submarine was Fulton's *Nautilus*, Figure 6.5. After attempting to gain support and finance in various European capitals, Napoleon Bonaparte finally provided Fulton with funds and *Nautilus* was built and launched on the Seine in 1801. She was built of copper on an iron frame and was 6.5 m long. Like the *Turtle* she submerged

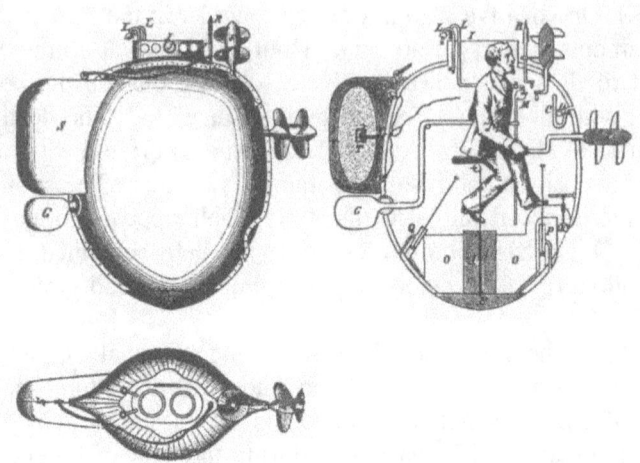

Figure 6.4 Bushnell's *Turtle*, 1776, after a drawing of 1885 in Kemp (1988, 193).

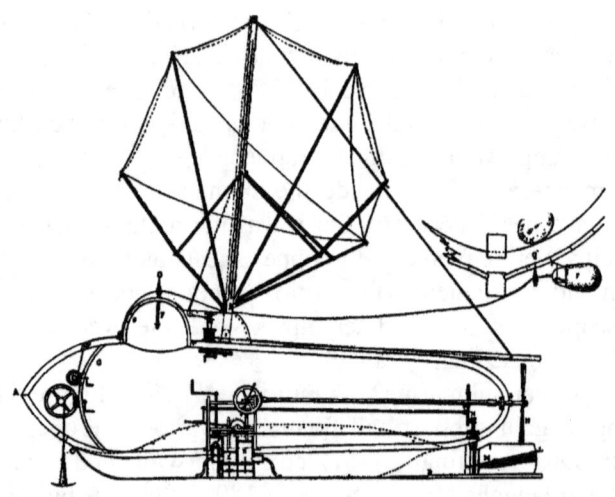

Figure 6.5 Robert Fulton's *Nautilus*, 1800. After Kemp (1988, 193).

by taking seawater into ballast tanks and was propelled under water by hand cranked propellers. On the surface, she could erect a mast and sail. She too attacked by approaching an enemy underwater and attaching an explosive to the enemy's hull, but by means of a spike rather than a screw. *Nautilus* proved to be very successful in trials but the French Navy, though impressed, could not consider such a terrible weapon and refused to proceed further. When Fulton demonstrated *Nautilus* to the British Navy they too considered it a diabolical weapon and one, moreover, that would lose them their command of the seas. Despite William Pitt's enthusiasm *Nautilus* was rejected and Fulton returned to America to become, eventually, a successful steamship operator.

Air Craft

The invention of a flying bird in the 4th century BC is described by Aulus Gellius (Attic Nights, 10.12.9-10), writing in the 2nd century AD. He says: "But the bird which the Pythagorean is reported to have designed and constructed certainly ought to seem no less astonishing, and not so fruitless. For not only many of the noble Greeks, but also the philosopher Favorinus, an avid student of old records, have written with the greatest assurance that, by certain principles and a knowledge of mechanics, Archytas made a wooden model of a dove that flew. Evidently it was balanced nicely with weights and propelled by compressed air within it. Concerning a story so difficult to believe, I prefer, by Hercules, to put down Favorinus' own words: 'Archytas of Tarentum, being a mechanical engineer, made a flying dove of wood. Whenever it alighted, it did not take off again. For until this---'[text break]"; Humphrey et al (1998, 62).

Gelius is right to be sceptical for it has long been man's dream to fly and history is littered with stories of flying machines and birdmen, mostly mythical, from Daedalus, through Shun, Ramesses, Bladud, Simon Magnus, Firman, Firnas, Djawhari, Eilmer, Leonardo, Danti, Damian and so on to Lilienthal.

Of these Eilmer, whose exploits took place about 1000 years ago, has some claim to be the earliest for whom hard evidence exists, see White (1978) reprinted in Woosnam (1986, 9). White points out that the evidence of Eilmer's flight is quite sound and comes from William of Malmesbury, a monk of Eilmer's own abbey, and was written barely 30 years after his death, when memories of the event were still fresh. It is evidence that the modern historian may accept. William of Malmesbury says that Eilmer "was a man learned in those times, of ripe old age, and in his early youth had hazarded a deed of remarkable boldness. He had by some means, I scarcely know what, fastened wings to his hands and feet so that, mistaking fable for truth, he might fly like Daedalus, and, collecting the breeze on the summit of a tower, he flew more than the distance of a furlong [201 m]. But agitated by the violence of the wind and the swirling air, as well as by the awareness of his rashness, he fell, broke his legs, and was lame ever after. He himself used to say that the cause of his failure was his forgetting to put a tail on the back part". White, based on the known facts of Eilmer's life, dates the flight to the first decade of the 11th century. It is believed that the tower at Malmesbury was then about 24 m high and to the north, west, and southwest there is a sharp drop of about 18 m to the river, so the prevailing southwest wind coming over the rise would give considerable lift. His glide angle, a 42 m fall in 201 m, is 12 degrees; see Figure 6.6.

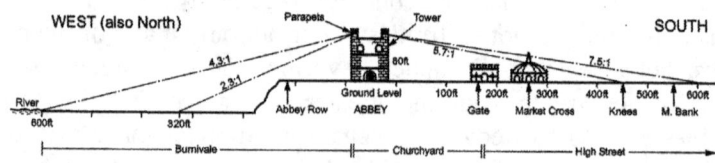

Figure 6.6 Possible glide angles from Malmesbury Abbey for Eilmer's flight of c1005. After Woosnam (1986, 24)

Leonardo da Vinci's interests and experiments with flight are now well known. In his attempts to attain manned flight he studied the flight of birds, air resistance, and sketched many devices for flapping wings by muscular exertion, but all to no avail. Some of his ideas, however, seem reasonable and his design for a parachute, Figure 6.7 (left) must be counted as successful although it is not known if it was ever tested in his lifetime. Leonardo says, "If a man has a tent 12 ells wide and 12 high covered with cloth he can throw himself down from any great height without hurting himself." It sounds quite precise and perhaps he had made measurements and trials, but without a hole in the centre for the air to escape it may have been quite unstable, eventually overturning. Or so it was always thought. Recently the parachute has been reconstructed to Leonardo's dimensions and it was found that it remained stable because air escaped through the weave of the cotton fabric. It has been used

Figure 6.7. Leonardo da Vinci's sketches, 1500.
Left: a parachute. Right: a balance for measure the lift of a wing.

successfully for a parachute descent from a hot-air balloon at about 3000 m.

Leonardo did attempt to measure the air resistance of wings, Figure 6.7 (right), although the device seems clumsy. He says "And if you wish to ascertain what weight will support this wing, place yourself on one side of a pair of balances and on the other place a corresponding weight, so that the two scales are level in the air; then if you fasten yourself to the lever where the wing is and cut the rope which keeps it up, you will see it suddenly fall; and if it requires two units of time to fall of itself, you will cause it to fall in one by taking hold of the lever with your hands; and you lend so much weight to the opposite arm of the balance that the two become equal in respect of that force; and whatever is the weight of the other balance, so much the wing will support as it flies; and so much it presses the air more vigorously."

Countless attempts at flight have been made throughout history. John Aubrey, in his brief life of Sir Jonas Moore says; "I remember Sir Jonas told us that a Jesuit (I think it was Grenbergerus, of the Roman College) found out a way of flying, and he made a youth perform it. Mr Gascoigne taught an Irish boy that way, and he flew over a river in Lancashire (or thereabouts) but when he was up in the air, the people gave a shout, whereat the boy being frightened, he fell down on the other side of the river, and broke his legs, and when he came to himself, he said that he thought the people had seen some strange apparition, which fancy amazed him. This was anno 1635, and he spake it in the Royal Society, upon the account of the flying at Paris, two years since", Lawson-Dick (1949, 210).

Real progress on manned flight had to wait until 1783 when the first hot-air balloon flights occurred. The Montgolfier brothers, Joseph and Étienne, were paper manufacturers at Annonay, near Lyon. Apparently, they noticed the way fragments of ash were carried upwards by a current of hot gas from a fire and this gave them the idea of capturing the hot gas in a bag to see if it would rise. They made a silk bag, filled it with hot fumes and found it would rise gently to the ceiling. By June 1783 they made a public demonstration of an unmanned balloon in Annonay. When members of the French Académie des Sciences heard of this they commissioned the physicist Jacque Alexandre Charles, of Charles' Law fame, to make a balloon. He made one of about 4 m diameter and filled it with hydrogen gas, which had been discovered in 1766. Charles' balloon was released in Paris in August 1783 to great acclaim and remained visible for two minutes before low cloud obscured it. The 4 m balloon came down at a little village, Gonesse, where it terrified the villagers, one of whom shot it with a gun, and when deflated it was torn to pieces.

Meanwhile, the Montgolfier brothers had moved to Paris where they had made a much larger balloon and after successful tethered tests it was decided to demonstrate it before Louis XVI and Marie Antoinette in September 1783. A sheep, a rooster and a duck were carried, suspended in a cage, and the flight covered about 2 km before returning to earth. The king was impressed and the Montgolfiers were elevated to the nobility. For the first flight with human passengers it was proposed that convicts should be sent up, but after some argument it was decided that criminals were not worthy of the honour of man's first flight and the physician Pilâtre de Rozier, a great supporter of the Montgolfiers, was chosen. His first tethered ascent was on 13th October 1783 when he ascended to a height of about 25 m. After several more trials the first free flight was made on 21st November 1783 with de Rozier as pilot and the Marquis d'Arlandes as passenger, see Figure 6.8 (left). After being filled with hot-air from a fire the balloon was released and floated for about 25 minutes over Paris, returning to earth, as the air cooled, after a flight of about 9 km. As early as 1797, Jacque Garnerin made a successful parachute drop from a hydrogen balloon, Figure 6.8 (right) but the parachute was not looked on as a means of escape in an emergency.

Figure 6.8 Left: The first untethered, manned flight of a hot-air balloon by de Rozier and the Marquis d'Arlandes in the Montgolfier brother's ballon.
Right: First parachute drop, by Jacque Garnerin, from a balloon in 1797.

Charles made another ascent in 1783, this time in an 8 m diameter hydrogen filled balloon in which, after a flight with his brother, he accidentally ascended, alone, to a height of about 2,700 m before returning safely to earth. He noted the low temperature at these altitudes and suffered extreme pain in his eardrums because of his rapid ascent. Although Charles never flew again his design of hydrogen balloon was to become the standard. A spherical envelope was of coloured, rubberised silk with a filling valve at the bottom and a release valve at the top, which the pilot could open using a cord to allow gas to escape when he wished to descend. The top of the balloon was covered with a net fastened to a circular hoop from which the passenger basket was suspended. Ballast, that could be thrown overboard to lighten the balloon, was carried.

The English Channel was crossed by wind driven balloon flight as early as 1785. Blanchard and Jeffries ascended at Dover, crossed the channel, and descended safely near Calais. Tethered balloons were used for reconnaissance during the Napoleonic wars by Guyton de Morveau who twice ascended during the battle of Fleurus, in 1794, to gather valuable information for Jourdain. Thousands of flights were made in the 19th century despite the numerous deaths and accidents that occurred. Ballooning became fashionable, and hydrogen, rather than hot-air, was preferred until late in the 20th century when improved heaters made hot-air cheaper, safer, and more convenient than hydrogen.

Bibliography

Adams, C., "Transport", in Scheidel, W., (Editor), "Roman Economy", Cambridge University Press, Cambridge, 2012.
Agricola, G., "De Re Metallica", Trans. Hoover, H.C., and Hoover, L.H., Dover, Publications, Inc., New York, 1950, originally published 1556.
Anon., "Wilton Windmill", Wiltshire County Council and Wilton Windmill Society, Trowbridge, Wilts, 1979.
Atkinson, F., "The Horse as a Source of Rotary Power", Trans. Newcomen Soc., 33, 31-55, 1960-61.
Aulius Gellius, "The Attic Nights", Reink Books, 2017. Also in Humphrey, J.W., Oleson, J.P., and Sherwood, A.N., "Greek and Roman Technology: A Sourcebook", Routledge, London, 1999.
Ausonius, "Mosella", in Humphrey, J.W., Oleson, J.P., and Sherwood, A.N., "Greek and Roman Technology: A Sourcebook", Routledge, London, 1999.
Bass, G.E. (Editor), "A History of Seafaring" Book Club Associates, London, 1974.
Bennett, R., and Elton, J., "History of Corn Milling", Volumes 1, 2, and 3, Simpkin, Marshall and Co. Ltd., London, 1898.
Biringuccio, V., "The Pirotechnia of Vannoccio Biringuccio", Translated and Edited by Smith, CS., and Gnudi, M.T., Dover Publications, New York, 1990, originally published 1540.
Borelli, G.A., "On the Movement of Animals", Translated Maquet, P, Springer Verlag, 1989. Published posthumously, 1710.
Bouger, "Traite du Navire de sa construction et de ses Mouvements", Paris, 1746.
Bowen, R.L., "Egypt's Earliest Sailing Ships", Antiquity, XXXIV, 117-31, 1960.
Branca, G., "Le Machine", Strenna Utet, Torino, 1977, originally published 1629.
Brett, G., "Byzantine Water Mill", Antiquity, XIII, 51, 354-356, 1939.
Brigg, T.H., "Haulage by Horses", Trans ASME, 14, 10 14-1065, 1893.
Brody, S., "Bioenergetics and Growth", Hafner Publishing Co. Inc., New York, 1968.
Brown, D.K., "Before the Ironclad", Conway Maritime Press Ltd., London, 1990.
Burford, A., "Heavy Transport in Classical Antiquity", Economic History Review, XIII, 1, 1-18, 1960.
Campbell, G.F., "China Tea Clippers", Adlard Coles Ltd., London, 1974.
Carus-Wilson, E.M., "An Industrial Revolution of the Thirteenth Century", Economic History Review, 12, 39-60, 1941.
Casson, L., "Ships and Seamanship in the Ancient World", Princeton University Press, 1971.
Casson, L., "Ships and Seafaring in Ancient Times", British Museum Press, London, 1994.
Chamberlayne, E., "Anglia notitia: or The Present State of Great Britain", London, 1669.
Cruden, S.H., "Click Mill, Orkney", Ancient Monuments and Historic Buildings, Ministry of Works, HMSO, Edinburgh, 1949.
Davidson, A.E., "Adventure", President's Address, Proc. I. Mech. E, 130, 33-369, 1935.

Davidson, C. St. C., "Transporting Sixty-Ton Statues in Early Assyria and Egypt", Technology and Culture, 2, 11—16, 1961.

Defoe, D., "A Tour Through the Whole Island of Great Britain", Penguin Books, 1971.

Denny, Sir Maurice, "BSRA Resistance Measurements on the Lucy Ashton", part 1, "Full Scale Measurements", Trans. Instn. of Naval Architects, 1951.

Derry, T.K., and Williams, T.I., "A Short History of Technology", Oxford University Press, Oxford, 1960.

Dickinson, H.W., and Straker, E., "The Shetland Watermill", Trans. Newcomen Society, XIII, 89-94, 1932-3.

Dickinson, H.W., "Origin and Manufacture of Wood Screws", Trans. Newcomen Soc., XXII, 79-85, l941-2.

Diderot, D., "A Diderot Pictorial Encyclopaedia of Trades and Industry", Edited by Charles Gillespie, Vols. 1 and 2, Dover Publications Inc., New York, 1959. Originally published 1751.

Diderot, D., and d'Alembert, I.J., "Encyclopèdie", Paris, 1772.

Diodorus of Sicily, "History", Transl. Oldfather, C.H., The Library of History, Harvard University Press, Cambridge, Mass. 1998.

Emerson, W. "The Principles of Mechanics", Reink Books, 2017, Re-printed from 1773 edition.

Evelyn, J., "The Diary of John Evelyn", Everyman's Library, No 221, Dent, Dutton, London, 1966.

Finley, M.I., "Technical Innovation and Economic Progress in the Ancient World", Economic History Review, 18, 29-45, 1965.

Foote, PG., and Wilson, D.M., "The Viking Achievement", Book Club Associates. London, 1979.

Forbes[1], R.J., "Studies in Ancient Technology", Vol. I, E.J. Brill, Leiden, 1955.

Forbes[2], R.J., "Studies in Ancient Technology", Vol. II, E.J. Brill, Leiden, 1955.

Forbes[3], R.J., "Studies in Ancient Technology", Vol. III, E.J. Brill, Leiden, 1955.

Froude, W., "On Experiments with H.M.S. Greyhound", Trans. Instn. of Naval Architects, 1874.

Gertler, M., "A Re-Analysis of the Original Test Data for the Taylor Standard Series", Navy Department, Washington DC, 1954.

Gillespie, C. (Ed.), "A Diderot Pictorial Encyclopaedia of Trades and Industry", Denis Diderot, Vols. 1 and 2, Dover Publications Inc., New York, 1959. Originally published in 1751.

Gimpell. J., "The Medieval Machine", Pimlico, London, 1993.

Glotz, G., "Un Transpot de Marbre pour le Portique d'Eleusis", Revue des ètudes Grecques, XXXVI, 26-45, 1923.

Glotz, G., "Ancient Greece at Work", Routledge and Kegan Paul Ltd., London, 1965.

Haden, W., "Letter to Board of Education", Wiltshire Record Office, 1325, 1928.

Hale, J.R., "The Lost Technology of Ancient Greek Rowing" Scientific American, 66-71, May 1966.

Harland, J., and Myers, M., "Seamanship in the Age of Sail", Conway Maritime Press, London, 1984.

Harris, D.G., "F.H. Chapman: The First Naval Architect and his Work", Conway Maritime Press, 1989.

Hart, I.B., "The World of Leonardo da Vinci", MacDonald, London, 1961
Hassall, W.O., "They Saw it Happen", Vol.1, 55 B.C. -1485, Basil Blackwell, Oxford, 1973. 1973.
Hassan, U, and Sykes, D.M., "Wind Structure and Statistics", in Ferris, L.L. (Editor), Wind Energy Conversion Systems, Prentice Hall, New York, 1990.
Herodotus, "The Histories", Translated de Sélincourt, A., Penguin Books, 1996.
Hill, A.V., "Muscles and Nerves in the Maintenance of Posture, the Development of Power, and the Transmission of Messages in the Body", Proc. I. Mech. E. 131, 333-81, 1935.
Hill, A.V., "The Heat of Shortening and the Dynamic Constants of a Muscle", Proc. Roy. Soc., Series B, London, 126, 136-195, 1938.
Hills, R.L. "Power from the Wind", Cambridge University Press, 1994.
Hockey, S.F. "Quarr Abbey and its Lands", Leicester University Press, 1970.
Hodgen, M.T., "Domesday Water Mills", Antiquity, XIII, 31, 261-279, 1939.
Hoerner, S.F. "Fluid-Dynamic Drag", Hoerner Fluid Dynamics, Albuquerque, NM, 1963.
Hogg, O.F.G., "English Artillery 1326-1716", Royal Artillery Institution, London, 1963.
Holmes, G., "The Later Middle Ages 1272-1485", Thomas Nelson & Sons Ltd., 1962.
Homer: "The Odyssey", Trans. Rieu, E.V., The Penguin Classics, London, 1946.
Hornell, J., "Water Transport: Origins and early Evolution", Cambridge University Press, 1946.
Humphrey, J.W., Oleson, J.P., and Sherwood, A.N., "Greek and Roman Technology: A Sourcebook", Routledge, London, 1999.
Jackman, W.T., "The Development of Transportation in Modern England", Frank Cass and Co., Ltd., 1962.
Jennings, H., "Pandaemonium", Papermac, London, 1995
Keil, I., "Building a Post Mill in 1342", Trans Newcomen Soc., XXXIV, 15, 1-4, 1962.
Keller, A.G., "A Theatre of Machines", Chapman and Hall, London, 1964.
Kemp, P., "The History of Ships", Black Cat, Macdonald and Co. (Publishers) Ltd., London, 1988.
Klemm, F., "A History of Western Technology", George Allen and Unwin Ltd., London, 1959.
Knight, E.H., "The Practical Dictionary of Mechanics", Volumes 1-5, Cassell & Co. Ltd., London, 1884.
Kuhn, H., Translated by Brodrick, A.H., "The Rock Pictures of Europe", Sidgwick and Jackson, London, 1956.
Lancaster Burne, E., Russel, T.J., and Wailes, R., "Windmill Sails", Trans. Newcomen Soc., XXIII, 147-161, 1943-5.
Landström, B., "Sailing Ships", George Allen and Unwin, London, 1969.
Lane, F.C., "The Economic Meaning of the Invention of the Compass", American Historical Review, 68, 605-617, 1963.
Lawson-Dick, O., "Aubrey's Brief Lives", Mandarin, London, 1949.
Lefebvre des Noëttes, RJ.E.C., "Le Chevalde Selle á Travers les Ages: Contribution á l'Histoire de l'Esclavage", Paris, 1931.

Leupold, J., "Theatrum Machinarum", 10 vols., Leipzig, 1724-39.
Littauer, M.A. and Crouwel, J.H., "Wheeled Vehicles and Ridden Animals in the Ancient Near East", E.J. Brill, Leiden, 1979.
Liversidge, J., "Britain in the Roman Empire", Book Club Associates, London, 1973.
Lopez, R.S., "The Evolution of Land Transport in the Middle Ages", Past and Present, 9, 17-29, 1956.
Lu, G.-D., Salmon, R.A. and Needham, J., "The Wheelwright's Art in Ancient China. 1. The Invention of Dishing", Physis (Florence), 1, 103-26, 1959.
Major, J.K., "The Horse Engine in the 19th Century", Newcomen Society Trans. 60, 31-48, 1988-89.
McCutcheon, W.A., "Water Power in the North of Ireland", Trans. Newcomen Society, 32, 67-94, 1966-7.
Morrison, J., and Coates, J., "The Athenian Trireme", Cambridge, 1986.
Morsley, C., "News from the English Countryside, 1750-1850", Harrap, London, 1979.
Muckle, W., and Taylor, D.A., "Muckle's Naval Architecture", Butterworths, London, 1987.
Ohlgren, I.H., (Editor), "Medieval Outlaws", Sutton Publishing, Stroud, 1998.
Parent, A., "Researches de Mathématiques et de la Physique", Paris, 1713.
Parsons, A.W., "A Roman Water-Mill in the Athenian Agora", Hispera, V, 70-90, 1936.
Pelham, R.A., "The Distribution of Early Fulling Mills in England and Wales". Geography, 13, 52-56, 1944.
Philipson, J., "The Art and Craft of Coachbuilding", George Bell and Sons, London, 1897.
Piggott, S., "The Earliest Wheeled Transport", Thames and Hudson, 1983.
Pliny the Elder, "Natural History", Translated Healy, J.F., Penguin Classics, London, 1991.
Postgate, J.N., "Early Mesopotamia", Routledge, London, 1992.
Pugh, R.B., "Court Rolls of the Wiltshire Manors of Adam de Stratton", Wiltshire Records Society, XXIV, Devizes, 1970.
Rahtz, P.A., "Medieval Milling", in Crossley, D.W., (Editor), "Medieval Industry", Research Report No. 40, Council for British Archaeology, 1-15, 1981.
Ramelli, A., "The Various and Ingenious Machines of Agostino Ramelli", Translated Gnudi, M.T., Annotated Ferguson, E.S., Dover Publications, New York, 1994. Originally published 1588.
Rankine, W.J.M., "Useful Rules and Tables Relating to Mensuration, Engineering, Structures, and Machines", Revised by Millar, W.J., Charles Griffin and Co., London, 1889.
Rieu, E.V. (Trans.), "Homer: The Odyssey", The Penguin Classics, London, 1946.
Ridley, A., "An Illustrated History of Transport", Heinemann, London, 1969.
Rostovtzeff, M., Translated by Duff, J.D., "A History of the Ancient World", Vols.1 and 2, HMSO and Oxford University Press, 1945.
Seppings, R., "On a New Principle of Constructing His Majesty's Ships of War", Phil. Trans. Royal Society, 285-302, 1814.
Singer, C., Holmyard, E.J., Hall, A.R., and Williams, T.I., "A History of Technology", Vol. II, c. 700B.C.-1500 A.D., Clarendon Press, Oxford, 1956.

Smeaton, J., "Experimental Enquiry Concerning the Natural Powers of Wind and Water to Turn Mills and other Machines Depending on a Circular Motion", I. and J. Taylor, London, 2nd Ed., 1796.
Stern, Van Doren, P., "Prehistoric Europe", Allen and Unwin Ltd., London, 1970.
Stowers, A., "Observations on the History of Water Power", Trans. Newcomen Society, XXX, 239-56, 1957.
Strabo, "The Geography of Strabo", Transl. Jones, H.L., Eight Vols, William Heinemann, Ltd, Cambridge, Mass., 1949.
Swanton, M., "The Deeds of Hereward", in Ohlgren, T.H., "Medieval Outlaws", Sutton Publishing, Stroud, 1998.
Templeton, W., "The Millwright and Engineer's Pocket Companion", Simpkin Marshall, London, 1856.
Thomas Smith, "The Complete Souldier", London, 1628.
Thrupp, G.A., "The History of Coaches", Kerby and Endean, London, 1877.
Thucydides, "The Complete Writings of Thucydides: The Peloponnesian War", Transl. Crawley, R., The Modern Library, New York, 1934.
Van Belle, R., "The World of Folly: The Foot-panels of the Walsokne Brass and the Persistence of their Iconography over the Centuries", Trans. Monumental Brass Soc., Vol XVII, part 3, 185-222, 2005.
Veranzio, F. "Machinae Novae", Venice, 1615.
Vitruvius, "On Architecture", X, 5, in Humphrey, J.W., Oleson, J.P., and Sherwood, A.N., "Greek and Roman Technology: A Sourcebook", Routledge, London, 1999.
Wailes, R., "Horizontal Windmills", Trans. Newcomen Society, XL, 125-145, 1967-8.
White, K.D., "Agricultural Implements of the Roman World", Cambridge University Press, Cambridge, 1967.
Wikander, O. (Editor), 'Handbook of Ancient Water Technology', Brill, Leiden, 377, 2000.
Wilson, P.N., "The Waterwheels of John Smeaton", Trans. Newcomen Society, XXX, 37-38, December 1955.
Wolf, A., "A History of Science, Technology, and Philosophy in the 16th and 17th Centuries", Vol. II, Peter Smith, Gloucester, Mass., 1968.
Wong, J.Y., "Terramechanics and Off-Road Vehicles", Elsevier, Amsterdam, 1989.
Wong, J.Y., "Theory of Ground Vehicles", John Wiley and Sons, Inc., New York, 1993.
Woosnam, M., "Eilmer, 11th Century Monk of Malmesbury: The Flight of the Comet", Friends of Malmesbury Abbey, Malmesbury, Wilts., 1986
Wright, T., "Thomas Young and Robert Seppings: Science and Ship Construction in the Early Nineteenth Century", Trans. Newcomen Soc., 53, 55-72, 1981.
Young, T., "Remarks on the Employment of Oblique Riders", Phil. Trans. (Royal Society), 303-336, 1814.
Xenophon, "Cyropaedia", in Humphrey, J.W., Oleson, J.P., and Sherwood, A.N., "Greek and Roman Technology: A Sourcebook", Routledge, London, 1999.
Zonca V., "Nova Teatro di Machine et Edificii", Padua, 1607.

Author Index

A
Adams, C., 11
Agricola, G., 2, 13
Alexander the Great, 121, 153
Antoinette, Marie, 162
Archimedes, 153
Aristotle, 122
Atkinson, F., 2, 13, 15
Attwood, George, 134
Aubrey, John, 161
Aulus Gellius, 159

B
Bacon, Francis, 60, 154
Bass, G. E., 122
Beaufoy, Colonel, 134
Beckman, 23
Bekker, 101, 105
Bennett, R., 13, 16, 23, 27, 29, 61
Bernoulli, 68
Besson, 52
Biringuccio, V., 4
Blanchard, 163
Blunt, Colonel, 89, 90, 95
Bouguer, 61, 133, 151
Bourne, William, 154
Bowen, R. L., 115
Brett, G., 33
Brigg, T. H., 99
Brody, 19, 20, 21
Brown, D. K., 133
Burford, A., 77, 82
Burne, Lancaster, 62
Bushnell, David, 157

C
Caliph Umar ibn al-Khattab, 51
Campbell, G. F., 135

Carus-Wilson, E. M., 35
Casson, L., 123
Cato, 85
Chamberlayne, E., 89
Chapman, F. H., 133, 136
Charles, Jacque Alexandre, 162, 163
Cicero, 85
Columbus, 131
Crouwel, 75

D
Daedalus, 160
Davidson, C., 78, 79, 80
Dickinson, H. W., 24
Diderot, D., 2
Diodorus, 122

E
Eilmer, 159
Elliot, Obadiah, 93, 95
Elton, J., 13, 16, 23, 27, 29, 61
Emerson, W., 61

F
Finley, M. I., 23
FitzHerbert, John, 37, 38
Forbes, R. J., 10, 51, 80
Froude, William, 139, 140
Fry, Elizabeth, 16
Fulton, Robert, 157

G
Garnerin, Jacque, 162
Gertler, M., 146
Gimpell, J., 9, 28
Glotz, G., 83, 123

H

Haden, W., 7
Halley, Edmund, 156
Harland, J., 152
Harris, D. G., 133
Hart, I. B., 57, 60
Haske, Thomas, 94
Hassall, W. O., 11
Hassan, U., 73
Heredotus, 119, 121
Hills, R. L., 53, 60, 65, 71t
Hilton, Robert, 63
Hockey, S. F., 35
Hodgen, M. T., 27
Hoerner, S. F., 150
Hogg, O. F. G., 87
Holmes, G., 86
Hooke, Robert, 1, 2
Hornell, J., 125
Humphrey, J. W., 9
Huygens, 133

J

Jackman, W. T., 87, 88, 91, 92
James I, 90
Jeffries, 163

K

Keller, A. G., 29
Kemp, P., 131, 154, 157
King William IV, 133
Klemm, F., 155
Knight, E. H., 29
Kühn, H., 125

L

Landström, B., 119, 123, 126, 129
Lawson-Dick, O., 162
Lee, Edmund, 63
Lefebvre des Noëttes, RJ. E. C., 9, 11, 82

Leonardo da Vinci, 153, 160, 161
Leupold, J., 2
Littauer, M. A., 75
Lopez, R. S., 85

M

McCutcheon, W. A., 37
Montgolfier, 162
Moore, Sir Jonas, 161
Morsley, C., 15
Myers, M., 152

N

Necho, 119
Neckham, Alexander, 153

O

Ohlgren, I. H., 87

P

Parsons, A. W., 31
Peregrinus, Peter, 129
Philipson, J., 93, 112
Piggott, S., 75, 76
Pilâtre de Rozier, 162
Pliny, 125
Prudentius, 33
Pugh, R. B., 53

Q

Queen Elizabeth, 87
Queen Hatshepsut, 117
Queen Mary, 87

R

Rahtz, P. A., 25, 28, 33
Ramelli, A., 2, 13, 28, 38, 39, 40, 56, 87
Rankine, WJ. M., 1, 3, 12, 42, 82, 109

Reynolds, 140, 141
Rippon, Walter, 87

S
Seppings, Robert, 134
Shaw, Captain, 99
Smeaton, John, 40, 41, 42, 62
Smith, Thomas, 87
Stowers, A., 28, 37, 41
Strabo, 23
Straker, 24
Sykes, 73

T
Templeton, W., 62
Thrupp, G. A., 85, 87

V
Van Belle, R., 56, 57
Verantius, 53
Veranzio, F., 53
Vitruvius, 39

W
Wailes, R., 52
Weir, 15
White, K. D., 159
Wikander, O., 26
Wilson, P. N., 40
Wong, J. Y., 105
Woosnam, M., 159

Z
Zonca, V., 2, 13

Subject Index

A
Adze, 76
agricultural revolution, 8, 9, 12
Alexandria, 51, 57
Anchor, 133
ancient civilisations, 3, 75–80, 115–121
Anglo-Saxon Chronicle, 27, 86
animal power, 1–21, 75, 80, 83, 85, 100, 101
Archimedean screw, 29
Archimedes, 153
Aristotle, 122, 153
Athens, 9, 31, 32, 121

B
Bacon, Francis, 60–62, 154
balance, 161
balloons, 153, 161, 162–164
band brake, 11
battle car, 76, 77
Bayeux tapestry, 12, 86, 88, 127
bearing, 4, 24, 29, 31, 35, 78, 107, 109, 130
Beaufort scale, 71
Belisarius, defence of Rome, 33
bellows, 4–5, 23, 37, 60, 62
bireme, 119
blacksmith, 4
block brake, 5, 7, 56
boats, 51, 115–119, 125, 126, 127–129, 130, 142, 143, 145, 146–148, 155
boring machine, 153, 154
bow wave, 139
bows of a cog, 130
brake, 5, 7, 56
band, 11
block, 31, 66, 82, 83, 117, 127, 144, 145
bronze, 78, 83, 122–123
Bronze Age, 80, 81, 125–126

C
cam, 37, 59
camel, 13, 80
canal, 117
canaries, 131
cannon, 80, 131–132
capstan, 9
caravan, 76, 117
caravel, 131
carbon, 122–123
carriage, 81, 83, 88, 90, 93, 94, 111–112
carriages, 81, 88, 90, 93, 111–112
carruca, 84, 85
cart, 9, 13, 75, 76, 78, 81, 82, 83, 85–87, 90, 91, 95, 96, 99, 100
cartwheel, 108
cast iron, 41, 63
casting, 41, 63
cattle, 75, 116
centrifugal fan, 63–64
chariot, 9, 10, 76, 77–79, 81, 82, 85, 90, 95, 97, 98
chiton, 8
Christianity, 9
Cistercian, 28
clippers, 134–136, 137, 145–146
clocks, 1
cloth, 35–37, 53, 66, 80, 81, 160
coaches, 85, 87–95
cog (ship), 130–131
colossi, stone, 3

column drums, 82
compass, 112, 129
composite bow, 1
copper, 123, 157
corn mill, 3, 7, 9, 13–14, 23–24, 25, 28, 30, 33, 35, 37–39, 51, 56, 63
crane, 6, 87
Crete, 26, 77
cross staff, 134
crown-and-lantern gear, 23, 31, 35, 53, 56, 59
Crusades, 53
C-spring carriage, 93, 94

D
Daedalus, 159, 160
Diderot's Encyclopaedia, 2
diving bell, 153, 154–156, 157
dog as power source, 13, 14, 15, 81, 118
draught animals, 76, 99–101, 102
drives, flexible, 11, 156

E
education, 1, 7
efficiency, 11, 15, 16, 20, 21, 25, 38, 40–49, 63, 66, 68–70, 75, 143, 147, 148
Egypt, 75, 77, 78, 79, 98, 115, 116, 119, 125
Eilmer, 159, 160
elliptical spring, 93–94, 95
endurance, 15, 16
energy, 1, 3, 15, 20–21, 26, 35, 40, 44, 47, 52, 60, 70, 107
engineer, 1, 15, 33, 40, 60, 62, 87, 117, 159
Etruscan, 120
Experiment (ship), 10, 13, 21, 57, 60, 62, 89, 90, 99, 105, 133–134, 138, 140, 142, 154, 160
explosion, 157, 159
explosive, 157, 159

F
fantail, 63, 64
farming, 75
fatigue, 35
feathering, 66–67
fire, 4, 13, 31, 57, 59, 80, 99, 131, 146, 162
fishhooks, 75
flight, 159, 160, 161, 162–164
floating mill, 33
forge hammer, 37
forging, 4, 37
fragile goods, 83
France, 28, 34, 90, 131, 133
friction, 3, 12–13, 39, 43, 44, 46, 48, 80, 99, 101, 102, 107, 109, 138, 140, 142, 146
frigate, 134, 135
Froude number, 134, 138–141, 142, 146
fulling, 35–37
fulling mill, 3, 23, 35–37
furnace, 4, 23, 37

G
Gaussian distribution of wind, 71
gear, 17, 31, 35, 44, 63, 110
gears, 13, 23, 24, 26, 29, 31, 33, 35, 44, 46, 52, 53, 56, 59, 63
gearwheel, 31
gin, 9
glass, 57, 88, 89, 90, 93, 153, 156
gold, 118
goods, 75, 83, 85, 86
governor, 7, 16, 63, 65, 66, 67
grain mill, 56, 63

granite, 117
Great Britain (ship), 89, 136
Great Pyramid, 116
Great Republic ship, 145–146
Greece, 81–85, 87, 98, 121–125
Greek mill, 23–30, 42, 46–49
gun, 131, 132, 133, 154, 155, 162
gunpowder, 3, 157

H

Hackney carriage, 88
hammer, 37
hard surfaces, 77, 98, 105, 107–109
harness, 9–12, 13, 77, 82, 83, 85, 87
heaving, 62
heels, 133, 134
Hereward the Wake, 87
hieroglyphic writing, 79, 117
hogging, 134, 143
horizontal-axis waterwheel, 31, 43, 45, 46, 47
horse, 2, 3, 9–14, 15, 19–21, 27, 56, 75, 77, 78, 80–89, 90, 91, 97, 99, 100, 101, 102, 110–112
horse harness, 9–12, 13
horse whim, 14
horse-driven machines, 13
horsepower, 15
horseshoes, 12, 82, 83, 87, 99, 101
hulls, clinker- and carvel-built, 126
Hungary, 87
hunting, 23, 80, 115

I

India, 75, 135
industrial revolution, 3, 24, 35
industrialisation, 7, 9, 153
Ireland, 26
iron, 4, 12, 35, 41, 63, 78, 83, 101, 112, 136, 144, 157

L

lamp, 153
land transport, 3, 75–113, 115
large-scale statues, 3, 4, 78, 80, 91, 117
Late Bronze Age, 81
lateen sails, 123–124, 131
lathe, 83
lead, 123, 125, 154, 155, 156, 157
leather, 57, 77, 78, 89, 90, 93
Leonardo, 57–58, 59, 60, 153, 154, 159, 160, 161
lever, 4, 24, 37, 161
lift pump, 57
load-carrying, 96, 112
locomotives, 13
long coach, 89
longitude, 134, 136, 143
lubrication, 3, 80

M

man power, 3–7
manorial system, 28
manual labour, 7
manufacture, 35, 37, 129, 162
mathematics, 61
measurement, 15, 19, 24, 42, 58, 96, 133, 141, 160, 161
mechanic, 1
mechanics, 9, 42, 153, 155, 159
Mediterranean shipbuilding, 115–116, 117, 119, 123, 124, 136, 143
merchant ships, 122, 123, 135
Mercury (newspaper), 15
Mesopotamia, 75, 76, 77, 80
metabolic rate, 15, 16, 19
metacentric height, 133
metal, 4, 12, 28, 38, 41, 77, 108, 136, 154
metal ores, 28

metal-working, 4
millpond, 24, 28, 39
mines, 23, 37, 125
monasteries, 28
monks, 28, 54, 159
mule, 1, 9, 83, 84
muscle, 1, 2, 3
muscle power, 3

N

Nautilus, 157, 158, 159
Navigation, 119, 131
Navy, 133, 159
Neolithic, 75, 80
Nile, 51, 115, 117, 119
Norman invasion, 127–128
Norse mill, 23–30, 42, 46–49
Nubia, 117

O

oar, 51, 115–116, 119, 121, 122, 123, 126, 127, 129, 141, 142, 146, 147, 148, 150
oil, 3, 85
onager, 76
ore, 23, 28, 37, 59
overshot wheels, 2, 31, 32, 34, 37, 38, 39, 40, 41, 42, 45, 46
oxen, 2, 3, 9, 11, 12, 13, 21, 27, 76, 82, 83, 85, 87, 91

P

Painting, 11, 57, 77, 116, 128, 152
Palaeolithic, 75
paper, 31, 60, 92, 133–134, 162
papyrus, 119
parachute, 160–161, 162, 163
patent, 52, 63, 64, 66, 67
paving, 87, 101, 109
Pelton wheel, 26, 42, 46–49
Perch, 93, 94, 113

Percussion, 61
Persia, 51, 80, 121
Phoenicians, 116, 119, 120, 121
pitting, 31, 58, 83, 117
plane, 61, 62, 130
Plough, 11, 76
ploughman, mediaeval, 11
Population, 35, 85
post mill, 12, 55, 56, 57
postal service, 94
posting, 56, 89, 93, 94, 115, 118
potter's wheel, 81
pottery, 81, 115, 120
power, 1–73, 75, 85, 97, 109–110, 111, 120, 138–143, 146, 147, 148, 150, 157
power source, 2, 3–21, 66
press, 9, 102, 161
pressure-sinkage relation, 102, 103
prime movers, 153
pump, 3, 7, 23, 37, 39, 51, 52, 56, 57, 61, 66, 155, 156, 157
Punt, 117, 118, 119
Pyramid, 3, 116, 117

Q

quadireme, 122
quadrant, 110, 133
quarrying, 82
Queen Hatshepsut, 117, 118
quern, 23, 26, 27, 28
quinquireme, 122

R

rag and chain, 52, 57
rail, 13, 92, 135
railway, 13, 92, 135
Renaissance workshops, 85–87, 125–132
Reynolds number, 140, 141, 142
ridging, 83

riding, 37, 78, 80–81, 87
road, 9, 12, 27, 28, 82, 83, 85–92, 95, 97, 99, 101, 105, 107–111
roller, 57, 79, 105
rolling, 13, 82, 96, 97, 98, 104, 105, 106, 107, 108, 109, 110, 112
Roman, 4, 8, 9, 13, 23, 24, 27, 30–41, 42–46, 49, 51, 83, 84, 85, 120, 121, 123, 124, 125, 126, 161
Roman Empire, 33, 121
Roman ship, 123, 124
Rome, ports, 82, 121
Rotary, 9, 13, 23
Rowing, 3, 27, 121, 146–150
Royal Mail, 95
Royal Society, 1, 40, 89, 90, 133, 134, 162
rubber, 163
rudder, 56, 123, 126, 129, 130, 157
running, 13, 16, 20, 89, 108

S

saddle, 77, 80, 82, 83
saddle stone, 23
safety valve, 58
sail, 3, 51, 52, 53, 56, 57, 60–64, 66–68, 70, 71, 115, 116, 118–131, 133–136, 141, 142, 146, 150–152
sailing, 51, 61, 115, 116, 121, 123, 124, 126, 129, 134, 135, 136, 140, 150–152
sailing ship, 51, 61, 115, 129, 134, 136, 140
sailing speed, 62, 66, 67
sailing, theory of, 60
Salamis, battle of, 121
saw, 16, 23, 76, 82
sawmill, 35

science, 133, 162
screw, 5, 29, 157, 159
screw (prison warder), 5
screw thread, 29
sea transport, 117, 119, 153–164
sea-going shipsSeneca, 51, 117
shell, 123, 124
ship, 28, 60, 86, 116–118, 121–124, 126, 127, 129–135, 138, 139, 141–145, 150, 152, 153, 154
shipbuilding, 115, 116, 117, 119, 123, 124, 136, 143
ships, 51, 61, 115–127, 129–136, 138–143, 144–145, 150, 152, 153–155, 157
shoe brake, 83, 84, 87
slavery, 7–9
sledge, 79, 80, 95
smelting, 4
Society for the improvement of Naval Architecture, 133
soft surfaces, wheels on, 101–106
soil mechanics, 98, 101, 102, 103, 105
Spain, 125
specialisation, Greek attitude to, 8, 11, 23–30, 42, 46–49, 51, 77, 81, 82, 97, 98, 120, 121, 159
speed governors, 7, 16, 63, 66, 67
spring, 66, 88, 89, 90, 93–95, 110, 112, 113
springing of vehicles, 95, 110, 112, 113
springs, elliptical, or laminated, 93, 94, 95
stability, 5, 36, 77, 133, 134, 160
Stamp mill, 58, 59
Standard of Ur, 76
Statue, 3, 4, 78, 80, 117

Steam, 13, 41, 57, 58, 59, 60, 65, 136, 159
Steel, 90, 93, 144
steering, 110–113, 115, 116, 119, 121, 123, 126, 129, 130, 157
stern wave, 139–140
stirrup, 82, 83, 86
stone, 3, 7, 8, 13, 23, 24, 26, 31, 35, 41, 44, 51, 52, 57, 63, 65, 78, 82, 83, 101, 109, 117, 125, 126
stone-crushers, 7
submarine, 153, 154, 155, 157
suspension, 31, 57, 85, 88, 89, 90, 93, 130, 153, 162, 163
Sweden, 125, 126, 133
Syria, 77, 121

T
technology, 8
tentering, 63, 65
Thames, 27, 88, 157
Theodosian Code, 9, 85, 90
Third Crusade, 53
tide mill, 28, 29, 53
tin, 125
tower mill, 26
traction, 99–101
traffic, 105
transport, 3, 9, 12, 51, 75–113, 115–164
treadmill, 2, 4, 5, 6–7, 13–15, 16–19
trip-hammer, 37
trireme, 121–122, 149, 150
turbine, 29, 53, 57, 58, 59
turnpike, 90, 95
tyre, 77, 78, 83, 91

U
undershot wheels, 2, 33, 37, 40, 42–46, 49
United States of America, 3, 65, 66, 75, 127, 135–136, 137, 145, 157, 159

V
Vasa (ship), 155
vehicles, 75, 76, 77, 83, 85, 86, 88, 90, 95–98, 99, 107, 110, 111, 112
vertical-axis waterheels, 23–30, 33, 46–49, 51–53, 59
Victor (ship), 86, 127
Viking longboat, 127
Vikings, 126, 127, 130

W
Wagon, 82, 84, 85, 90, 91, 110
warship, 119, 120, 121, 122, 123, 130, 131, 132, 143, 155, 157
water mill, 1, 23, 25, 27, 28, 31, 33, 34, 42, 45, 53, 55
water mills, 3, 23, 27, 28, 33, 35, 37, 42, 53, 55
water supply, 31, 37
water transport, 3, 75, 115–152
waterpower, 35
waterwheel, 2, 28, 29, 30, 31, 34, 35, 37, 38, 39, 40, 41, 42, 43, 44, 45, 49
Weibull distribution of wind speed, 70–71

Weir, 15
well, 3, 7, 11, 41, 51, 56, 57, 136
wheel, 2, 5, 7, 13, 23, 24, 26, 28–31, 33–35, 37–49, 75–78, 81, 83, 85–86, 88, 90, 91, 95–98, 101–113
wheelbarrow, 95, 98
whim, 9, 13, 14
wind, 3, 9–10, 11, 51–73, 119, 122, 123, 124, 126, 131, 133, 150, 152, 160, 163
wind power, 2, 3, 51–73
wind rose, 130
windmill, 1, 12, 13, 51–70
window, 52, 57, 88, 89, 90, 156
worms, marine boring, 157

www.ingramcontent.com/pod-product-compliance
Lightning Source LLC
Chambersburg PA
CBHW051744230426
43670CB00012B/2161